animals&men

THE JOURNAL OF THE CENTRE FOR FORTEAN ZOOLOGY

Issue 49

Typeset by Jonathan Downes,
Cover and Layout by SPiderKaT for CFZ Communications
Using Microsoft Word 2000, Microsoft , Publisher 2000, Adobe Photoshop CS.

First published in Great Britain by CFZ Press

CFZ Press
Myrtle Cottage
Woolsery
Bideford
North Devon
EX39 5QR

ISBN: 978-1-905723-78-2

EDITORIAL

Dear Friends,

It is time for a change. I have been editing *Animals & Men* for over seventeen years, and before I go any further let me get one thing straight; I intend to carry on editing it for so long as I am able. I have no intention of closing the title.

However, as regular readers are aware, things have been a little uncertain as far as our publishing schedule has been concerned for the last few years. I will be the first to admit that it was my fault; I totally miscalculated the production costs of the magazine now that it is produced perfect bound, but the real reason is that we are now doing so much more than we used to.

This year alone so far we have published nine books (with another five which will almost certainly have been published by the time that you read this), and we are just about to finish our second feature-length film *Emily and the Big Cats*.

I have been biting off more than I can chew for most of my life, and I have no intention of stopping now.

However, it is becoming very evident that (as I said at the top of the page) it is time for a change. Since 2002 a subscription to *Animals & Men* has been synonymous with membership of the Centre for Fortean Zoology. I think that this probably should change.

We are looking to completely change the membership scheme to become a yearly production, with a substantial discount on both ALL books published by the CFZ, plus Weird Weekend tickets. A&M will no longer be a direct part of the yearly membership, but you will get a discount whenever it comes out, if you are a member (the same applies for TAN).

A quarterly e-newsletter will also be sent out to members, keeping subscribers up to date with the latest goings in the CFZ, including: big news stories, recent publications and the latest news about the CFZ.

So, to be clear, the old membership subscription got you 4 issues of A&M (or TAN) and was renewed every 4 issues.

The new membership will be paid for every year, but you will get a discount on all books published by the CFZ, all periodicals published by the CFZ, discounted tickets to the WW, and a quarterly e-newsletter.

"THE GREAT DAYS OF ZOOLOGY ARE NOT DONE"

On the other hand, the fact that this publication comes out so irregularly does not really affect your subscription which is for four issues rather than for twelve months, so everyone gets what they pay for in the end.

It is up to you. What do you think?

Please let me know, in confidence if you prefer, and I promise that I will take everybody's views into account when I make my final decision…

Changing the subject completely, regular readers of the daily CFZ blogs will know that last November, the CFZ expedition to India brought back samples from some antlers found in India.

The full expedition report (available in book form from CFZ Press) tells the story:

> Llewellyn, a conservationist rather than a hunter, invited us to look at his father's collection. Eagle-eyed Jon McGowan spotted something unusual among them - a pair of muntjac horns of unbelievable size.

> On closer examination, these very distinctive horns proved to be even larger than those of the giant muntjac *(Muntiacus vuquangensis)* of Vietnam and Laos. This picture shows the horns next to those of the Indian muntjac *(Muntiacus muntjak)*, and the startling size difference is apparent. Local people have a name for this particular deer, calling it 'matchok'. We took some samples from the antler for analysis back in Europe.

We sent the samples to Lars Thomas, and he passed them on to Prof. Tom Gilbert at Copenhagen University. For a while it looked as though our suspicions were justified, but after hours of painstaking work, Lars eventually wrote to me:

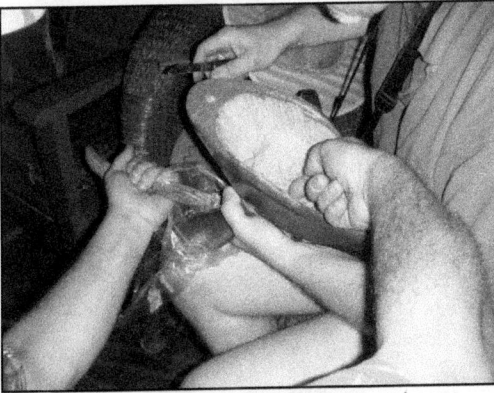

"Hi Guys, Finally we have the results of the DNA analysis of the antler samples from India. It has taken an awful lot of time, but we do need to check and recheck and check again - and earn a living every now and then :-). Unfortunately there was no new species in there after all... "

The antlers turned out to be from a Sambhur (Sambar) deer - (*Rusa unicolor*) presumably a juvenile - because although the antlers of an adult resemble those of a fallow deer, the antlers of the juvenile surprised me greatly by looking like those of a muntjac. It is mildly disappointing but our job is to find out the truth, and not purely to look for cryptids, and we have found the truth.

Richard Freeman and I would like to thank Tom Gilbert and his team for all their painstaking work. And as I often do, I am going to take refuge in the words of Rudyard Kipling:

As the dawn was breaking the Sambhur belled --
Once, twice and again!
And a doe leaped up, and a doe leaped up
From the pond in the wood where the wild deer sup.
This I, scouting alone, beheld,
Once, twice, and again!

Changing the subject yet again:

As regular readers or visitors to the CFZ itself will know, two years ago we acquired eight very young specimens of an undescribed species of cichlid - Aequidens spp - from Peru. We lost three of them in a power cut during the winter of 2009/2010 and another one a couple of months ago in a fight between the two largest specimens.

However, in early July we were happy to announce that they have bred.

When they breed again, we will be happy to share them to any aquariums or private fish-keepers (especially those with experience of cichlids, and either a species tank or a collection of large cichlids)

We shall be studying the growth of these little fishes closely, as next to nothing is known about them, and we hope, soon, to be able to identify them once and for all, and to publish the results..

As well as the cichlids we have bred another 12 species of fish so far this year, but our most exciting breeding success are the two baby Rio Cauca caecilians that are presently ambling around the big show tank in the conservatory.

These little amphibians are obscure enough already, but to breed them is very rare - in the UK anyway.

We obtained ours at about the same time as the ZSL and Durrell Wildlife, and I am feeling rather smug to announce that to the best of my knowledge we bred ours first.

Things are looking up for the CFZ at the moment. Let's hope that it stays that way.

Onwards and Upwards

Jon Downes
Director, Centre for Fortean Zoology.

THE FACULTY OF THE CENTRE FOR FORTEAN ZOOLOGY

"In her abnormalities, nature reveals her secrets." (Goethe)

PERMANENT DIRECTORATE

Hon. Life President:
Colonel John Blashford-Snell

******

Director:
Jonathan Downes
Deputy Director:
Graham Inglis
Administrative Director:
Corinna Downes
Zoological Director:
Richard Freeman
Deputy Zoological Director:
Max Blake
Technical Director:
David Braund-Phillips
Expedition Co-ordinator
Adam Davies
Marketing
Glen Vaudrey/Rebecca Lang

******

Ecologist: Oll Lewis
Directorate Assistant: Liz Clancy

Big Cat Study Group:
Neil Arnold
Jonathan McGowan
Aquatic Monster Study Group:
Oll Lewis
BHM Study Group:
Paul Vella
Bigfoot Forums
Ken Osburn

*****

American Office:
Nicholas Redfern
Naomi West
Richie West
Canadian Office
Robin Pyatt Bellamy
Australian Office:
Rebecca Lang and Mike Williams
New Zealand Office:
Tony Lucas

Surrey:
Nick Smith
Tyneside:
Mike Hallowell
West Midlands:
Dr Karl Shuker
Raheel Mughal
Wiltshire:
Matthew Williams
Yorkshire:
Steve Jones
Mark Martin

Wales

Gavin Lloyd Wilson
Oll Lewis
Gwilym Ganes

Northern Ireland

Gary Cunningham
Ronan Coghlan

USA

California:
Greg Bishop
Dianne Hamann
Illinois:
Jessica Dardeen
Derek Grebner
Indiana:
Elizabeth Clem
Michigan:
Raven Meindel
Missouri:
Kenn Thomas
Lanette Baker
New York State:
Peter Robbins
Brian Gaugler
New Jersey:

Brian Gaugler
North Carolina:
Shane Lea
Micah Hanks
Ohio:
Chris Kraska
Brian Parsons
Oklahoma:
Melissa Miller
Oregon:
Regan Lee
Texas:
Chester Moore
Naomi West
Richie West
Ken Gerhard
Nick Redfern
Wisconsin:
Felinda Bullock

International

Australia:
Tim the Yowie Man
Mike Williams
Paul Cropper
Rebecca Lang
Tony Healey
Denmark:
Lars Thomas
Eire:
Tony 'Doc' Shiels
Mark Lingard
France:
Francois de Sarre
Germany:
Wolfgang Schmidt
New Zealand:
Tony Lucas
Roo Besley
Peter Hassall
Switzerland:
Georges Massey
United Arab Emirates:
Heather Mikhail

CONTENTS

The Centre for Fortean Zoology is a non-profit making organisation administered by CFZ Natural History Ltd, (Company #7381545) a Company Limited by Guarantee.

From humble beginnings we have become the largest, and we like to think the best, cryptozoological organisation in the English-speaking world. We have published over 85 books, 49 editions of our journal *Animals & Men*, as of October 2011, 48 episodes of our monthly webTV show *On the Track*, 4 feature length films and dozens of shorter features, 12 annual conferences and over a dozen foreign expeditions.

We still hope to become a charity. With a track record like this, what can we hope to achieve by attaining charitable status? Surely it is not just about the money. Well of course, it is not *just* about the money, but money is a necessary evil. Whilst we are – we hope – justifiably proud of our achievements to date, if we manage to get extra cash in as a result of our attaining charitable status there are a number of projects that we want to fund. We would like to provide laptops and software for various researchers who would otherwise not be able to afford them. We want to set up static educational displays featuring posters, models and live exhibits in schools, hospitals, hospices, pubs, and restaurants across the southwest, to expand our programme of educational activities, and we would like to widen the scope of publishing projects. So keep your fingers crossed.

EDITED AND COMPILED BY CORINNA DOWNES

Welcome to the *Animals & Men #49* Newsfile. Once again time has flown by rather too quickly for my liking, and it has taken me quite a while looking back at all the new and rediscovered species that have been reported since the last issue. As ever, it is amazing that so many creatures are still being found, or re-found, as the case may be. It is nice to see some larger species too this issue.

So often these pages seem to be biased towards smaller creatures that can so easily be missed in their natural habitats, so to be able to include such 'beefy' animals as the Bornean bay cat has been a refreshing change.

For such large varieties of life to suddenly be re-discovered is a marvel and serves to instil encouragement that there are other such animals out there still to be found, hiding behind leaf and twig, or blending in with their natural environment so well that our eyes have not yet detected them. I hope you enjoy reading the following selection.

Iran's endangered cheetahs are a unique subspecies

Genetic experts have revealed that Iran's critically endangered cheetahs are the last remaining survivors of a unique ancient Asian subspecies, *Acinonyx jubatus venaticus*. DNA comparisons show that these Asiatic cheetahs split from other cheetahs, which live in Africa, 30,000 years ago and researchers have suggested that Iran's cheetahs must be conserved in order to protect the future of all cheetahs. Formerly existing in 44 countries in Africa, cheetahs are now only found in 29. Historically, however, they were also recorded across southwest and central Asia, but can now only be found in Iran.

Scientists have theorised that approximately 10,000 years ago there was a 'population crash', which led to inbreeding of the species suggesting that cheetahs have low genetic variability. Scientists from the University of Veterinary Medicine in Vienna, Austria have been working in collaboration with the Iranian Department of Environment and wildcat conservation group *Panthera* with the hope

NEW & REDISCOVERED

that they can solve the puzzle of modern cheetahs' origins.

Dr Pamela Burger and her team analysed the DNA of cheetahs from a wide geographical and historical range, including medieval remains found in north-western Iran, and explained:

> "With our data we prove that current Iranian cheetahs represent the historical Asiatic subspecies *A.j. venaticus* as they share a similar genetic profile with specimen originating from northwestern Iran in 800-900 CE."

It is currently estimated that there are just 60-100 individuals, with less than half at mature breeding age, and the Iranian cheetah population is classified as critically endangered by the IUCN Red List. The Iranian Department of the Environment has established a programme together with the United Nations Development Programme, Panthera and the Wildlife Conservation Society in order to make conservation of the Asiatic cheetah a national priority.

SOURCE: 24 January 2011
By Ella Davies, *Earth News reporter*
http://news.bbc.co.uk/earth/hi/earth_news/
newsid_9365000/9365567.stm

Two new species of 'leaping' beetles discovered in New Caledonia

After a three-year study Spanish researchers have discovered two new herbivorous beetles (*Arsipoda geographica* and *Arsipoda rostrata*) in New Caledonia in the western Pacific. Previously only five species of these so-called 'flea' beetles – out of a worldwide total of 60 – had been found there. However, interestingly, these new beetles have presented a mystery: they feed on plants that the scientists have yet to find on the archipelago.

Jesús Gómez-Zurita is lead author of the study and a researcher at the Institute of Evolutionary Biology (CSIC-UPF). He is also passionate about New Caledonia and has collected hundreds of beetles in order to study them. He told SINC:

> "The study, financed by the National Geographic, went some way beyond merely classifying species, and investigated the ecology of these herbivorous insects with a prodigious leaping capacity, which they use to avoid their predators."

In order to classify the DNA sequences of the animals' diet, in particular chloroplast DNA (which is exclusive to plants) the researchers - from Spain and New Caledonia - used previously-developed molecular tools. They used plant matter remains found in the digestive tract of the insects at the time they were killed in order to extract their DNA at the same time. This technique made it possible to dis-

cover that one of the new species, *Arsipoda geographica*, which measures three millimetres, feeds on a tropical plant in the mountains (Myrsinaceae), while *Arsipoda isola*, which is the same size, feeds on another plant (Ericaceae) in the southern jungles of the island.

SOURCES: Gómez-Zurita, J., Cardoso, A., Jurado-Rivera, J.A., Jolivet, P., Cazères, S. Mille, C. 2010. "Discovery of new species of New Caledonian

Arsipoda Erichson, 1842 (*Coleoptera: Chrysomelidae*) and insights on their ecology and evolution using DNA markers" *Journal of Natural History* 44(41-42): 2557-2579, 2010.
18 January 2011
http://www.eurekalert.org/pub_releases/2011-01/f-sf-tns011811.php

Elusive Cat Once Thought Extinct is Alive and OK in Borneo

Nearly all that is known by scientists about the Bornean bay cat comes from only 12 samples. The first of these was a skin collected in 1855 in Sarawak, Malaysia and in the several decades following this, seven other skins appeared.

However, it was not until 1992 that a living specimen was obtained, followed by another capture in 1998, after which the species seemed to disappear. Widely thought to be extinct, it was not until 2003 that a photograph was taken of a single cat. Since then photographic evidence has extended to include a further three photographs which depicted two or three individuals taken by camera traps in 2010. The research officer overseeing the camera trap project, Wilhelmina Cluny, explained:

> "This species is very secretive...it was classified as extinct until a photograph of it was taken in 2003...I do feel encouraged, this photograph was taken in a logged forest...when we saw this it made us wonder whether this kind of habitat can sustain wild-

life, even for rare and important species like the bay cat."

SOURCE: David DeFranza, Washington, DC on 01.14.11 Travel & Nature
http://www.treehugger.com/files/2011/01/elusive-cat-once-thought-extinct-is-alive-and-ok-in-borneo.php?camaign=th_rss&utm_source=feedburner&utm_medium=feed&utm_campaign=Feed%3A+treehuggersite+%28Treehugger%

Scientists Discover New Fish Species

By using modern genetic analysis and tradiional examination of morphology, scientists at the Smithsonian Institute and the Ocean Science Foundation have discovered that there are 10 distinct species of blenny in the genus Starksia rather than the three that were once thought. The Starksia blennies are small at less than 2 inches, and have elongated bodies.

They are generally native to the western Atlantic to eastern Pacific oceans where they live in shallow to moderately deep rock and coral reefs.

They have been studied for more than 100 years and it was thought there was little left to discover about them, although the utilisation of modern DNA barcoding techniques has proved those ideas wrong.

SOURCE: http://www.redorbit.com/news/
science/1991596/
scientists_discover_new_fish_species/index.html

Cryptic new wolf species identified in north Africa

The cryptic African wolf: *Canis aureus lupaster* is not a golden jackal - new molecular evi-

dence has revealed that there is a new species of grey wolf living in Africa. Formerly confused with golden jackals, and thought to be an Egyptian subspecies of jackal, the new African wolf shows that members of the grey wolf lineage reached Africa about 3 million years ago, before they spread throughout the northern hemisphere.

In 1880, evolutionary biologist, Thomas Huxley said that Egyptian golden jackals (then, as now, regarded as a subspecies of the golden jackal) looked suspiciously like grey wolves.

Several 20th Century biologists studying skulls made the same observation, but the conventional taxonomy has not been changed.

A collaboration of biologists from the University of Oslo, Oxford University's Wildlife Conservation research Unit (WildCRU) and Addis Ababa University has undertaken a new study and has discovered genetic evidence that places the Egyptian jackal within the grey wolf species complex.

It is a wolf, not a jackal, and is taxonomically grouped with the Holarctic grey wolf, the Indian wolf and the Himalayan wolf.

Dr Eli Rueness, the first author of the paper, states,

> "We could hardly believe our own eyes when we found wolf DNA that did not match anything in GenBank."

One of the authors of the paper - Professor Nils Chr. Stenseth - (who is also the Chair of the Centre for Ecological and Evolutionary Synthesis (CEES)) said,

> "This study shows the strengths of modern genetic techniques: old puzzles can be solved."

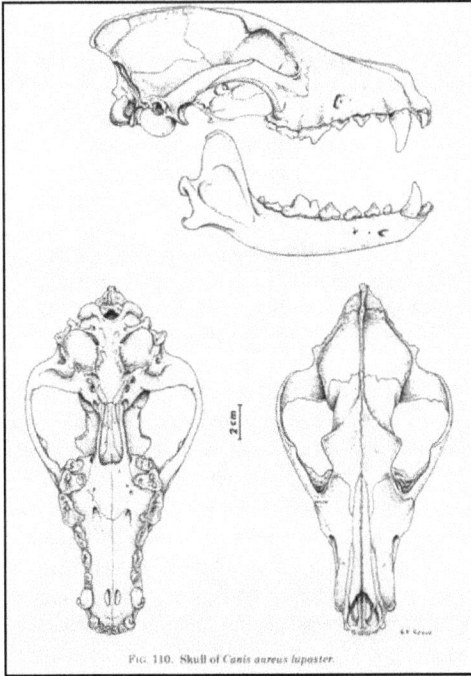

Fig. 110. Skull of *Canis aureus lupaster*.

Professor Afework Bekele at Addis Ababa University concluded:

"This shows how genetic techniques may expose hidden biodiversity in a relatively unexplored country like Ethiopia."

SOURCE: January 2011
If you are interested in more pictures and location of this exciting observation, please
click: www.greeneye.org.uk
http://www.wildlifeextra.com/go/news/egyptian-wolf.html

Scientists have identified a new type of mosquito which has raised concern

Researchers have told *Science* magazine that the new mosquito appears to be very susceptible to the parasite that causes malaria. It may well have evaded classification before now as it tends to rest away from human dwellings where most scientific collections tend to be

made. It is a subgroup of *Anopheles gambiae*, the insect species responsible for most of the malaria transmission in Africa.

Over a period of four years, Dr Michelle Riehle, from the Pasteur Institute in Paris, France, and colleagues gathered mosquitoes from ponds and puddles near villages and made their discovery. After examination of the insects in the lab, it was found that they are genetically distinct from any *Anopheles gambiae* insects previous recorded. Generations of the unique subtype were grown in the lab by the team so as to aid their assessment of their susceptibility to the malaria parasite and it was found that they were especially vulnerable.

However, Dr Ken Vernick, Pasteur team-member, cautioned that these mosquitoes' significance for malaria transmission had yet to be established. He said:

"We are in a zone where we need to do some footwork in the field to identify a means to capture the wild adults of the outdoor-resting sub-group. Then we can test them and measure their level of infection with malaria, and then we can put a number on how much - if any - of the actual malaria transmission this outdoor-resting subgroup is responsible for."

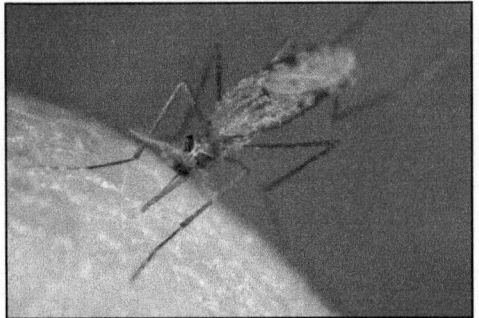

The researchers report that the new subgroup could be quite a recent development in mosquito evolution and urge further investigation to understand better the consequences for malaria control, and commenting on the study, Dr Gareth Lycett - a malaria researcher from the Liverpool School of Tropical Medicine in the UK - said it was an interesting advance that might have important implications for tackling malaria.

He told BBC News:

"To control malaria in an area you need to know what mosquitoes are passing on the disease in that district, and to do that you need sampling methods that record all significant disease vectors. You need to determine what they feed on, when and where, and whether they are infectious. And where non-house-resting mosquitoes are contributing to disease transmission, devise effective control methods that will complement bed-net usage and house spraying. A recent 12m-euro multi-national project (AvecNET), funded by the European Union, and led by the Liverpool School of Tropical Medicine has the specific aims of doing just this."

According to the World Health Organization (WHO), there are more than 200 million cases of malaria worldwide each year, resulting in hundreds of thousands of deaths, most of them in Africa.

Malaria is caused by Plasmodium parasites. The parasites are spread to people through the bites of infected female Anopheles mosquitoes.

SOURCE: 3 February 2011
By Jonathan Amos Science correspondent, BBC News
Source: http://www.bbc.co.uk/news/science-environment-12352565

New Bat in the Caribbean

The Caribbean island of St. Vincent has had a new species of bat declared by researchers. Although this new bat has been previously documented, it had been thought to have been a member of a similar species found throughout South America and a few of the Caribbean islands. However, PhD student, Peter Larsen, noticed that it was larger than its southern relative.

As Peter Larsen (who is studying at Texas Tech University) said in a press release:

"A year or so went by [after collecting the species] and I happened to look at this species [...] and compared it to what we thought it was - a species from Trinidad. But the St. Vincent bat was huge comparatively speaking."

According to Larsen when he spoke to mongabay.com (as source below), the bat is "a few grams heavier and is a few millimetres longer in most measurements that we took" than the big-eared bat (*Micronycteris megalotis*) – its closest relative. These differences may not sound a lot, but it is worthwhile remembering that a bat's weight is around 8 grams and measures around 30 millimetres, making a few grams and millimetres quite a big difference.

Compared to the nine other known species of Micronycteris bats, this particular species is 'medium-sized' and researchers believe that the new bat species may have become stranded from the mainland population be-

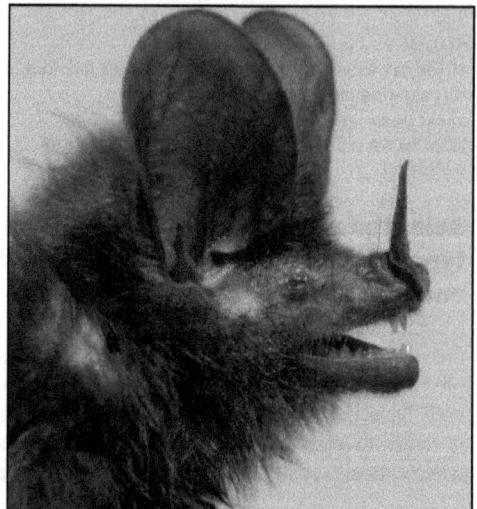

tween 600,000 and a million years ago. Larsen and his team decided to name the species the Garifuna big-eared bat (*Micronycteris garifuna*) after the Garifuna people who inhabit St. Vincent and other areas of the Caribbean and Central America. Its prey is insects thus providing an ecosystem service to the island. Another graduate student working on the study, Lizette Siles, said in a press release:

"They can actually pick their insect prey off the surface of rocks and leaves. Not all insectivores can do that, because most insectivores catch their prey on the fly. Their big ears, wide wings and membranes between the rear feet and tail allow them to maneuver better."

SOURCE: Peter A. Larsen, Lizette Siles , Scott C. Pedersen, Gary G. Kwiecinski. A new species of Micronycteris (Chiroptera: Phyllostomidae) from Saint Vincent, Lesser Antilles. Mammalian Biology (2011). doi:10.1016/j.mambio.2011.01.006. Jeremy Hance, mongabay.com, May 26, 2011 http://news.mongabay.com/2011/0526-hance_newbat.html

New bee discovered with largest tongue in world

A new species of bee has been discovered in the southern Colombian department of Nariño by scientists at Colombia's National University (UN). According to the monthly university publication, *UNPeriodico*, *Euglossa natesi* n. sp. - also known as the "orchid bee" or "jewel bee," - has a tongue twice the size of its body. Recognised by scientists for its luminescent mix of blue, green, bronze and gold, as well as its abnormally large tongue, this new type of jewel bee was found near the Ecuadorian border in southern Nariño.

"This insect is unusual, because it has the largest tongue found thus far and measures two times the size of its body," said professor Rudolfo Ospina, the director of the biology department at UN. According to the professor, the large tongue is used to reach nectar in orchids that cannot be reached by other species of bees, thus allowing the pollination of different types of orchids. The bee is plentiful in the lowland areas of the Neotropics, which is an area of similar flora and fauna that extends from Mexico covering the majority of Central and South America. Ospina said: "It is possible that some species also live in dry and open habitats".

Euglossa natesi n. sp. is part of the Euglossa genus and was named in honour of UN professor Guiomar Nates for her contribution to the research of bees.

Source: http://colombiareports.com/colombia-news-lite/news/16937-new-bee-discovered-with-largest-tongue-in-world.html Original report (in Spanish):
http://www.unperiodico.unal.edu.co/dper/article/nueva-especie-de-abeja-con-lengua-descomunal.html

Red rodent shows up at Colombian nature lodge after 113 years

The red-crested tree rat (*Santamartamys rufodorsalis*) - or the red crested soft-furred spiny-rat - had not been recorded since 1898 and was thought to be more than likely extinct. However, that all changed when one was seen at 9:30 pm on May 4th at a lodge in El Dorado Nature Reserve in northern Colombia. It was said to be about the size of a guinea pig.

"He just shuffled up the handrail near where we were sitting and seemed totally unperturbed by all the excitement he was causing,"

...said Lizzie Noble, a British volunteer with Fundación ProAves a conservation group in Colombia focusing on birds. The creature had only been known from two

skins previously, but the species was odd enough to be given its own genus, Santamartamys. Researchers are expecting the species will be listed by the IUCN Red List as Critically Endangered; indeed the species is believed to be imperilled by feral cats already.

Lina Daza, Executive Director of ProAves, said:

"We are so proud that our El Dorado Nature Reserve has provided a safe haven for this enigmatic little guy to survive. The discovery exemplifies why we buy forested properties known to be important for endangered wildlife yet at imminent risk of being destroyed."

SOURCE: Jeremy Hance
mongabay.com
May 18, 2011
http://news.mongabay.com/2011/0518-hance_redtreerat.html

Lost rainbow toad is rediscovered

A team of scientists - led by Dr Indraneil Das - from the Universiti Malaysia Sarawak (UNIMAS) spent months in the remote mountain forests of the Gunung Penrissen range of Western Sarawak (a boundary between Malay-sia's Sarawak State and Indonesia's Kalimantan Barat Province) and have rediscovered a species of toad that was long thought extinct. They found three colourful long-legged Borneo rainbow toads up a tree whilst searching at night – a species that had previously only been shown in illustrations of specimens collected by European explorers in the 1920s. Dr Das said:

"Thrilling discoveries like this beautiful toad, and the critical importance of amphibians to healthy ecosystems, are what fuel us to keep searching for lost species. They remind us that nature still holds precious secrets that we are still uncovering."

Dr Robin Moore of Conservation International, which launched its Global Search for Lost Amphibians in 2010, said:

"To see the first pictures of a species that has been lost for almost 90 years defies belief. It is good to know that nature can surprise us when we are close to giving up hope, especially amidst our planet's escalating extinction crisis. Amphibians are at the forefront of this tragedy, so I hope that these unique species serve as flagships for conservation, inspiring pride and hope by Malaysians and people everywhere."

Conservation International had listed this toad as one of the "world's top 10 most wanted frogs".

Source: http://www.bbc.co.uk/nature/14151541

New Species of Desert Tortoise

According to a new study co-authored by USGS the desert tortoise is comprised of two distinct species – the newly recognised species being named Morafka's desert tortoise (*Gopherus morafkai*) and most tortoises from Arizona and Mexico will now be assigned to this species.

Some complicated genetic analysis was needed to sort out the two species, which is not really surprising - considering the secretive animal involved has adapted to merge perfectly with the desert back ground and keep out of sight.

A co-author of this study, Kristin Berry who is a USGS Western Ecological Research Center biologist said that desert tortoises were first known to science 150 years ago. They were described by army doctor and scientist James Graham Coopoer in 1861 and were given the scientific name *Gopherus agassizii* and all

populations within the known range (California, Nevada, Utah, Arizona and Mexico) were believed to belong to that species.

Kristin Berry said:

"Populations on opposite sides of the Colorado River had different habitat preferences. Tortoises south and east of the Colorado prefer to hide and burrow under rock crevices on steep, rocky hillsides, while tortoises north and west of the Colorado prefers to dig burrows in valleys.

There were other minute differences as well, such as egg-laying seasons and little details in the tortoises' shell - differences you couldn't tell from a casual glance at these leathery clawed crawlers."

These differences led Berry and colleagues Bob Murphy, Taylor Edwards, Alan Leviton, Amy Lathrop and Daren Riedle into organising a little hunt, which involved sorting through and digging up hundred-year-old pickled specimens of desert tortoises from the vaults of the Smithsonian and California Academy of Sciences. They then utilised complicated DNA analysis to compare tortoise genes from the ancient, preserved tissue with samples from living tortoises from throughout the desert southwest.

SOURCE: USGS Desert Tortoise News, Tuesday, June 28, 2011
Written by: Ben Young Landis

Inflatable Shark Among 300 New Species Discovered in Philippines

Hundreds of new species seem to have been discovered in the Philippines, including a sea star that feeds completely on sunken driftwood and a deep-sea, shrimp-eating shark that swells up to scare off other predators. More than 300 species which are probably new to science have been uncovered from the

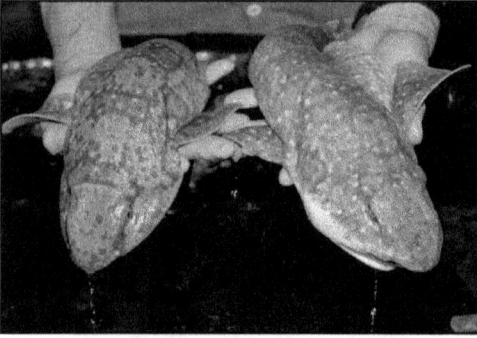

mountains to the sea by scientists, including dozens of insects and spiders, sea slugs and deep-sea armoured corals.

Researchers from the California Academy of Sciences and colleagues from the University of the Philippines and the National Museum of the Philippines undertook a 42-day expedition to survey Luzon Island, which is the largest island in the Philippine archipelago, and its surrounding waters.

Terrence Gosliner, dean of science and research collections at the California Academy of Sciences and leader of the 2011 Philippine Biodiversity Expedition said:

"We had our work both on the coral reefs and rain forest interrupted by an early typhoon; we were out of the water for two days."

He continued:

"One of the biologists working in the mountains was sleeping in a hammock; during the night, one of the trees his hammock was tied to was uprooted and he was suddenly on the ground,"

Discoveries include: a cicada with a distinctive 'laughing' call, a crab whose pincers are lined with needlelike teeth, and a wormlike pipefish that hides among colonies of soft coral. In addition, they discovered a possible new species of swell shark - a shark that pumps water into its stomach to puff up - which unlike its relatives possesses a very distinctive camouflaged colour pattern. A lot of the species inhabit places that are rarely, if ever, visited by people, such as a primitive plant called a spikemoss from the perilously steep upper slopes of Mount Isarog and a snake eel from the bottom of the ocean. Many others have avoided detection in the past because of their diminutive size, such as goblin spiders and barnacles that all measure just a few millimeters long.

Gosliner said that all these new findings help support the idea that the Philippines "is one of the hottest of the hotspots for diverse and threatened life on Earth. We found new species during nearly every dive and hike as we surveyed the country's reefs, rainforests and the ocean floor".

Source: http://beta.news.yahoo.com/inflatable-shark-among-300-species-discovered-philippines-135007930.html
By Charles Q. Choi , LiveScience Contributor
LiveScience.com *Jun 27, 2011*

Two rare Indian pheasants' new territory

The shy and silent western tragopan is extremely rare, but in June 2011 it was recorded at two new sites along the Pir Panjal range in Jammu and Kashmir. Sightings and calls of the pheasant were validated at the Kalamund-Tatakuti and Khara Rakh areas of the range. It is a Schedule I species on the Indian Wildlife (Protection) Act and listed as 'Vulnerable' by the IUCN Red List.

The western tragopan is a medium-sized, brightly coloured pheasant endemic to the western Himalayas and inhabits coniferous forests. Locals had talked about seeing the bird in April - but its presence was confirmed the following month. "The bird is extremely shy and silent. But knowing that the best way to locate the species would be during its

breeding season, when it becomes highly vocal, we returned in May," said Riyaz Ahmad, the team leader and assistant manager, species division of WTI.

The bird has become a victim of extensive poaching for both its meat and plumage, and together with habitat degredation and fragmentation, it had only been previously reported from the Kazinag range and Kishtawar National Park in the state. There have been a few scattered records occurring from the Sud Mahadeo area of Jammu province.

> "I was pleasantly surprised to note the tragopan's presence in these areas. Unlike its usual haunts, the moist north-facing coniferous slopes, the present sites are located on the south face of Pir Panjal along Poonch,"

...said Dr Rahul Kaul, South Asia representative, IUCN SSC Galliformes Specialist Group and Chief Ecologist, WTI.

As well as the western tragopan, the team also sighted another threatened species in the region, the cheer pheasant. The team has recommended Kalamund-Tatakuti for notification as a protected area as it is ecologically diverse and they believe it is representative of western Himalayan forests and it possesses key species such as the markhor, brown bear and musk deer.

SOURCE: http://www.wildlifeextra.com/go/news/indian-pheasant.html June 2011

New species of marine snail found off Florida

A new marine species of nudibranch - *Chromodoris fentoni* has been found in the Gulf of Mexico by biologists with the Florida Fish and Wildlife Conservation Commission (FWC), and scientists from California State Polytechnic University. This nudibranch was first observed in 2009 after sponges and other specimens were donated to the FWC's Fish and Wildlife Research Institute (FWRI) in St Petersburgh by Daniel Felton – a commercial aquarium-trade fisherman. He had collected these specimens from the Gulf of Mexico – off Tarpon Springs – and whilst sorting through them, biologists from the FWRI, Nancy Sheridan and Joan Herrera, saw the unusual creature.

> "We were unable to identify one of the nudibranchs and realized that it was possible we were seeing something entirely new,"

...said Sheridan.

> "The discovery was especially rewarding because it resulted from a cooperative effort between industry and science."

The sent samples to Dr Angel Valdes of California State Polytechnic University and they were verified as never having been previously documented.

"The opportunity to work with Dr. Valdes, a world-class nudibranch expert, has been really exciting for us,"

...said FWRI Curator of Collections, Dr. Joan Herrera.

"At FWRI, we receive thousands of specimens each year, yet it is rare to find a species that is new to science."

Adult nudibranchs are a member of the phylum Mollusca and have no external gills or shell. They feed on sponges, corals, anemones and other sea life and come in various shapes and sizes, ranging from 1/8-inch to 2 feet in length.

They are colourful creatures with bright red markings on an off-white background and have oblong bodies, which reach approximately 1 inch in length. April 2011

SOURCE: http://www.worldfishingnetwork.com/news/fish-catch-reveals-new-species-85263.aspx

A caddisfly species has been found for the first time in the UK

Stuart Crofts discovered a new caddisfly species near a small stream flowing through woodlands near Masham in North Yorkshire in January 2011.

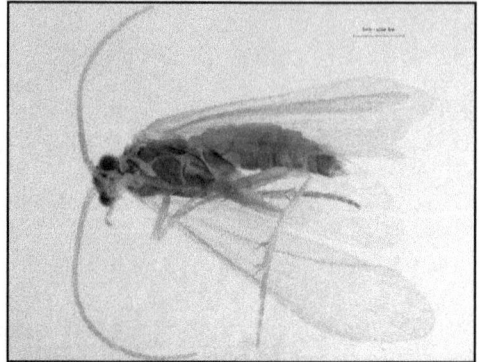

The last time a new species of caddisfly was found in the UK was in 1965. They spend most of their life as larvae in freshwater before emerging as adult flies and are fascinating insects. They are pollution sensitive and are commonly used as a litmus test for the health of the environment. They aid healthy bird and fish populations due to the fact that they provide an important food source.

"I couldn't believe what I was seeing when I identified it"

...said Stuart Crofts, a former international fly fisherman and Coordinator of the Adult Caddisfly Occurrence Scheme.

"To find a species new to the UK is amazing, a great honour and very humbling experience".

Synagapetus dubitans, is a small caddisfly of around 5mm and is more commonly found in central Europe

SOURCE: January 2011
Courtesy of The Riverfly Partnership, Riverfly Recording Schemes and Buglife - The Invertebrate Conservation Trust
http://www.wildlifeextra.com/go/news/caddisfly.html

OBITUARY

Williiam Corliss passed away 8 July, a month and a half shy of his eighty-fifth year. Corliss, for those who don't know, was the world's greatest living anomalist. From 1974 to his death he collected curiosities culled from science magazines and journals. He then took these anomalies and ordered them into categories ranging from Astronomy through Zoology, categories that were then published in whole or in sub-categories. These publications included an assessment of the anomaly in question, possible explanations, some astute observations and typically a number of exclamation marks. Many of these works are already worth a lot of money – Beachcombing has been plucking up the courage to buy Ancient Infrastructure (1999). Others will now become so as a belated tribute to WC.

Comparisons with Charles Fort (obit 1932) are often made but are surely misplaced as it is not so much the similarities between the two men as the differences that matter. Fort was a visionary and, despite his denials, knew it. Corliss had a sense of humour that only the non-committed can enjoy. Fort took reports wherever he would find them. Corliss

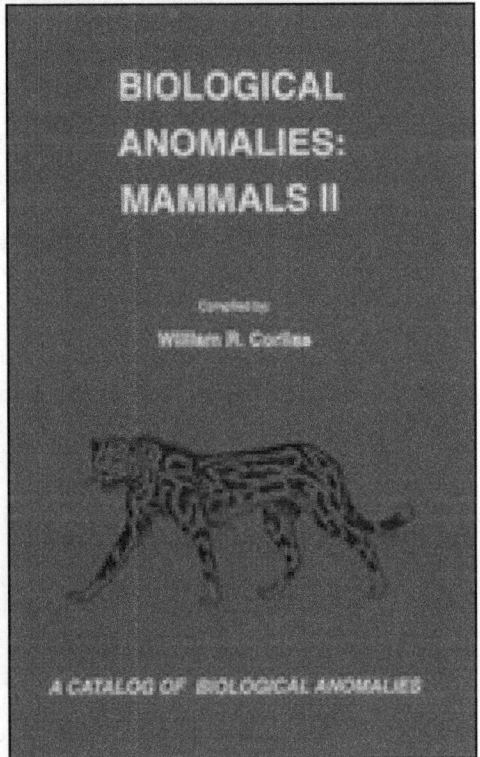

BIOLOGICAL ANOMALIES: MAMMALS II

Compiled by
William R. Corliss

A CATALOG OF BIOLOGICAL ANOMALIES

tended to restrict his searches to academically accredited works. Fort was on the soft end of the humanities with prose to match, Corliss was a scientist with remarkable

WILLIAM CORLISS
1926-2011

range and a usefully bland style. Fort was a one man Punch and Judy show who published five books and attracted disciples: 'Forteans swarmed to him like settlers, he became a land'. Corliss created a system of anomaly collection that transcended him and that will hopefully survive his death. It would be absurd to talk of Corlissians.

Beachcombing is fond of both CF and WC, though he feels far more at ease with Corliss's scientific and 'secular' approach. He certainly feels that this death concerns him more than CF's mysterious passing on the best part of a century ago. Beach gives this belated pledge of loyalty, in part, because he spent most of this spring reading Corliss while waiting in the half light for his elder daughter to drop off. The tiny rubrics required only intermittent concentration: just as well as the reading was constantly interrupted by objections to barbie doll privileges and concerns over monsters under the bed. Beach had even worked up a short list of queries and suggestions for additions that he had wanted to send in to the great man, but was put off by the fact that he only had a postal address and no email. Another regret that will fester like an unrecoverable splinter…

Commiserations to the Corliss family and WC's many friends. **Dr Beachcombing**

TURA SATANA, (1938-2011)

We all have certain experiences that have a profound effect on us. One of mine is when, as a teenager, late night on Channel 4, I first watched Russ Mayer's cult exploitation movie *Faster Pussycat Kill! Kill!* The plot involved three go-go dancing, leather clad, ultra violent girls, who, in search of thrills, go on a spree of killing, robbery and seduction. John Walters

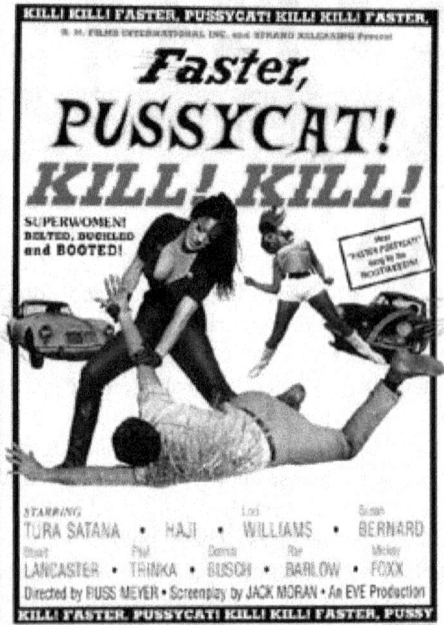

once described it as 'the best movie ever made'.

Leader of the three violent femmes was Varla, a raven haired, voluptuous killing machine with the cruellest and most beautiful eyes imaginable. She was played by the actress Tura Satana whose real life was even more amazing than any of her screen rolls.

Her real name was Tura Luna Pascual Yamaguchi. She was born on July 10, 1938 in Hokkaido, Japan. Her father was a silent movie actor of Japanese and Filipino descent. Her mother was half Cheyenne Indian and half Scots-Irish. The mix gave Tura her unique and unforgettable looks.

Her family moved to the USA but with the Japanese feelings during the war they were forced to live in an internment camp. African American girls tried to bully Tura at school but she soundly beat them. She developed quite young and was horrifically gang-raped on her way home from school by a gang of youths. After this she took up martial arts and these would serve her well for the rest of her days. The youths who raped her were never brought to justice on account of a bribed judge. Tura was sent to a reform school and at the age of just 13 had an arranged marriage to 17 year old John Satana.

After all this you can't blame the girl for developing a rebellious streak. She ran away a year after her marriage and ran away to Los Angeles with a fake ID and ended up modelling nude for Harold Lloyd (yes that Harold Lloyd). After a while she returned to Chicago to live with her parents and started dancing. She became a successful erotic dancer, earning about $1,500 per week.

Her peerless beauty secured her many acting rolls in TV during the 60s including the *Man from U.N.C.L.E, Burke's Law, Hawaiian Eye,* and *The Greatest Show on Earth.* Soon she was in demand on the big screen and appeared in such films as *Doll Squad, Irma la Douce, and The Astro-Zombies* (the latter being an amazing b-movie about a mad scientist who, having been fired by the space agency, decides to create superhuman monsters from the body parts of innocent murder victims, as you do. The creatures of course go on the rampage.)

She dated Elvis Presley for a while and turned down his proposal of marriage. She finally wed a lucky LA policeman Endel Jurman in 1981 and remained with him till his death in 2000. She left acting in the 1970s and became a nurse. During this time she was shot by a former lover who was high on drugs. In the 80s she was involved in a car accident that broke her back. She spent the next two years in and out of hospital, having a total of 17 operations. She was told that she would never walk again but typically for her she refused to accept this and learned to walk again.

Recently she had returned to acting including a sequel to *Astro-Zombies* entitled *Mark of the Astro-Zombies.*

Last year she became a friend of mine on Facebook. I found her to be a delightful lady, now 72 but showing little signs of slowing down. She told me about her early life and I was amazed that she was not bitter about all the awful things that had befallen her. She told me that those things had made her into a stronger person, the person she was today.

Tura passed away on February 4th in Reno, Nevada of heart failure.

Goodnight sweet Tura. The world will not know your like again. **Richard Freeman**

NICHOLAS COURTNEY 1929-2011

I've always said that *Doctor Who* and in particular the Jon Pertwee years were what inspired me to become a cryptozoologist. I will never forget those Saturday evenings when, for half an hour between 'a round up of the regional news and sport' and *The Generation Game* I would be transported into another world, a world inhabited by giant killer maggots, Sea Devils, Autons, Daleks, Axons, Silurians and other fantastical horrors. Besides the Doctor himself one man was always there, UNIT's Brigadier

Alistair Lethbridge-Stewart, ably played by the wonderful Nicholas Courtney.

Nick was born in Cairo, a diplomat's son and was educated in a number of places including Kenya, France and Egypt. He could speak Arabic and French before he could speak English. After his national service he joined the Webber Douglas Academy of Drama were he was awarded the Margaret Rutherford medal. Later he did repertory theatre in Northampton before moving to London.

During the 1960s he appeared in a number of classic shows such as *The Avengers, The Champions* and *Randall and Hopkirk (Deceased)*. In 1965 he was cast in the *Doctor Who* story *The Daleks' Masterplan* staring William Hartnell as the first Doctor. His portrayal of Space Security Agent Bret Vyon so impressed director Douglas Camfield that he cast him in two later stories *The Web of Fear* (the one with robot yetis on the London underground and an alien intelligence that manifests as an ecoplasmic web) and *Invasion* (featuring the Cybermen haunting the sewers beneath London) both with the second Doctor, Patrick Troughton. These stories introduced United Nations Intelligence Taskforce or UNIT (basically the army's alien

fighting division) and Colonel Alistair Lethbridge-Stewart.

When the third, and best Doctor arrived in the shape of Jon Pertwee in 1970s *Spearhead from Space* the Colonel had become a Brigadier. For the next five years he appeared alongside Jon Pertwee and then Tom Baker battling such foes as the Loch Ness Monster (a saurian cyborg created by the alien Zygons from their crippled spaceship beneath the Loch's waters), alien daemons, and rogue time lords The Master and Omega.

The dynamic between the actors during this era has never been replicated and is largely due to Nick's wonderfully pompous Brigadier at odds with Jon's authority hating Doctor. The show tackled some difficult themes. In the 1970 adventure The Silurians the Brigadier bombs the cave system were the intelligent race of reptiles, the titular Silurians, former rulers of the earth, are hibernating. The Doctor talked them into returning into suspended animation to avoid conflict with humans whom they see as a pest much like rats.

The Brigadier's appearances became less with the advent of Tom Baker as the fourth Doctor and a return to more stories based in outer space.

The Brigadier returned in the 1983 stories *Mawdryn Undead* and *The Five Doctors* and the woeful 1989 effort *Battlefield*.

In the revamped series UNIT has returned but seemingly plays second fiddle to the pointless and irritating *Torchwood*. The Brigadier was mentioned several times but did not actually appear. Nick's took up the roll one last time in the spin off series *The*

Sarah Jane Adventures. He was said to have been on a mission in Peru for sometime. He remarked that he disliked the new way UNIT works. Sadly ill health prevented him from reprising the roll. That was a real shame as I had always wanted to see the Brigadier return to Doctor Who and give the insufferable prats at Torchwood a good hiding.

After his main stint on Doctor Who Nick featured in a number of shows such as *Minder*, *Only Fools and Horses*, and *Yes Prime Minister*.

I was lucky enough to meet him several times and found him to be utterly charming. Nick died on 22 February 2011 at the age of 81 after a long battle with illness. One hopes that on reaching the Pearly Gates he said "Archangel Gabriel? Chap with wings, five rounds rapid!" **Richard Freeman**

ELISABETH SLADEN 1946-2011

As a kid at my school, we played at Dr Who *en-masse*. I was usually the third Doctor or sometimes the fourth. Other kids would play the Brigadier, Benton, Yates or other members of UNIT. Other kids would be Sea Devils, Autons, Yeti, Axons (spaghetti men, we called them), Daleks, Zygons, Krynoids or Cybermen. But the girls; they all wanted to be Sarah Jane Smith.

Sarah was the companion to many people. She was beside the Doctor in many of his most memorable adventures. Never just a yes-girl or a helpless shrinking violet, Sarah was often just as instrumental in saving the day as the Doctor himself. When she came to the role, Liverpudlian actress Elisabeth Sladen was relatively unknown, having had small roles in *Some Mothers Do Ave Em*, *Doomwatch*,

Coronation Street and *Z-Cars*.

When the brilliant Katy Manning bowed out of *Dr Who, Z-Cars* producer Ron Craddock recommended Elisabeth to Berry Letts who was *Dr Who*'s producer at the time. She proved an instant hit with both Letts and Jon Pertwee.

Her character was one of the most fondly remembered companions in the history of the series. Especially fondly remembered is the team of Tom Baker, Liz Sladen and the late Ian Marter as the fourth Doctor, Sarah and Harry Sullivan.

Such was her popularity that the character was brought back time and again, firstly in 1981 for the ill-fated spin off *K9 And Company*, then in 1983 for the 20th anniversary special *The Five Doctors*.

With *Dr Who*'s triumphant return Sarah Jane teamed up with David Tennant in the story *School Reunion*. Finally - perhaps inevitably - Sarah Jane got her own series,

The Sarah Jane Adventures, aimed at younger viewers, but still intelligently written and well acted. She was brought back for a showdown with Davros and the Daleks in *Stolen Earth* and *Journey's End,* again with David Tennant. She had last encountered them in the 1975 epic *Genesis of the Daleks.* Davros even comments on remembering her face. Liz worked alongside the new Doctor, Matt Smith, and her predecessor Jo Grant in *Death of the Doctor.*

I met Katy Manning shortly after and she eulogised on what a joy Liz had been to work with. It was great for many of the old fans who recalled Sarah Jane as a heart throb from the '70s to see her back, now like an eccentric, lovable favourite aunt. Liz died suddenly on April 19th. She had been suffering from cancer but few fans knew; she looked so well and had carried on with her acting regardless. It feels as if a part of the universe has fallen away leaving a bottomless void. **Richard Freeman**

LARRY "WILD MAN" FISHER (1944-2011)

Even within the CFZ office there are massively mixed opinions about the legacy of Wild Man Fisher, and these are nothing to the mixed opinions of him in the wider world.

He was diagnosed with both Paranoid Schizophrenia and Bipolar Disorder (despite the fact that various experts including my own psychiatrist have claimed that the two disorders are mutually exclusive. As a Bipolar sufferer (manic depressive) myself I have often felt a strange attraction to the recorded work of the Wild Man, and have never accepted the PC viewpoint that he was just exploited by his one-time mentor Frank Zappa. It is certain that when Zappa released Fisher's debut LP

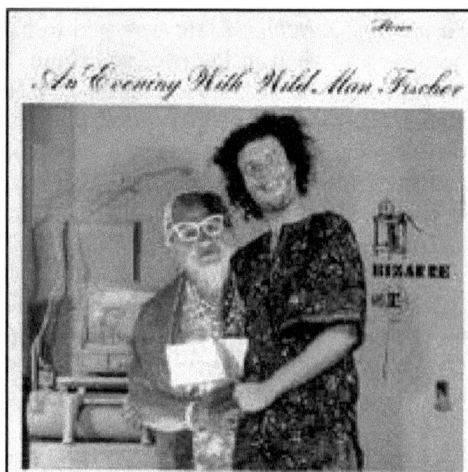

(see above) in 1968 as part of his programme to prove that *anyone* could be a pop star. At the time, Fisher (who had been hospitalised after attacking his mother with a knife at the age of sixteen) was a homeless man in Los Angeles who eked out a pitiful living busking and extemporising songs for passers-by at ten cents a throw.

He became a celebrity of sorts with appearances on *Rowan and Martin's Laugh In* but after his much publicised falling out with Zappa did little until several albums with avant-rockers *Barnes and Barnes* (one of whom was Bill Mumy, star of *Lost in Space* and *Babylon 5*). His tortuous relationship with Mumy and others is shown in an unputdownable VDV *Derailroaded* (2005). In the wake of this rediscovery he finally started taking his medication, and according to friends and family was never the same again. You may ask why there is an obituary of "The Godfather of Outsider Music" in a magazine primarily about cryptozoology, but if you have to ask that question it would appear that you don't know very much about the CFZ. **Jon Downes**

NICK REDFERN'S
Letters from America

Dogmen, Werewolves and Full-Moon Monsters

I f you have followed the written output of Linda Godfrey and her work on werewolf-style entities seen in the United States, then you'll definitely want to grab a copy of her brand new book: *The Michigan Dogman: Werewolves and Other Unknown Canines Across the U.S.A.*

Linda has carved for herself a well-deserved reputation as the leading U.S.-based investigator, researcher and author - in a non-fiction setting - of data on such matters, as is evidenced by her previous titles, *The Beast of Bray Road*; *Hunting the American Werewolf*; and *Werewolves*.

And, *The Michigan Dogman* is a fine addition to Linda's previous titles. You might possibly wonder: what can be said about such things that haven't already been said in Linda's previous three books on werewolves and lycanthropy? Well, the answer is quite a lot!

One of the chief reasons why *Dogman* is such an important and captivating read is because we're treated to countless new cases, eye-witness reports, and incidents and testimony - from all across the United States, and the decades, too.

And this is a significant factor: whereas *The Beast of Bray Road* was very much a regional study of werewolves in one particular part of Wisconsin, Linda's new book skillfully demonstrates that, in reality, these things - whatever they may be - have been seen, and are still being seen, all across the United States. In other words, this is not a localized, regional phenomenon. Rather, there appears to be something among us that has carefully avoided classification and capture for...well, who knows how long? But, it's also something that pops us just about here, there and everywhere, and appears to be somehow intimately connected with us - which I'll expand upon shortly.

The sheer range and variety of reports makes *The Michigan Dogman* essential reading for devotees of hairy, fanged werewolves. But, as interesting as the reports, are the notable similarities in the actions, characteristics and appearances of the creatures under scrutiny.

In other words, people from all across the United States - unconnected to, and unknown to, each other - are reporting sightings of what sound very much like the exact same creatures. Typically, they are large, hair-covered, possess muzzles and pointed ears, and have the ability to move on both two legs and four.

Indeed, the sheer number of reports that possess all of these particular aspects is astonishing. And unless hundreds of people are all banding together to hoax Linda - which I do not, for one moment, believe - then we have to address seriously the idea that there are creatures out there that look like the classic imagery of werewolves.

But, is that really what they are? Well, like all of Linda's books, she gets into some fascinating areas of research as she strives to answer that question, which some might assume is a simple one to answer, but I assure you it's most definitely not!

Linda, to her credit, does not shy away from controversy - and when it comes to the Michigan Dogman and its motley ilk, there's plenty of it! Without doubt the most important questions of all are: What are these creatures? Are they even flesh-and-blood, physical entities? Is something far stranger afoot? Are we looking at several phenomena that have been lumped together under one banner?

The curious thing about the Dogmen, as Linda carefully demonstrates, is that they seem to defy categorization. Aside from the fact that there is not - or certainly should not be - any sort of canine animal running around the United States that has the ability to walk, run and leap on two legs as effortlessly as it does on four, this is what the witnesses are reporting. And, even though such a scenario is controversial in the extreme, this still seems to place these creatures in a physical, flesh-and-blood, category - as does the fact that many witnesses have seen such beasts feeding on their prey near the sides of wood-shrouded roads late at night. But, this is where it gets even more interesting. It's almost as if there are too many cases: it's

clear from reading *The Michigan Dogman* that, on many occasions, the very fact that these creatures are seen goes beyond mere chance. It's almost like they want to be seen - or, perhaps, even need to be seen. Why? Well, we'll get to that soon.

In the same way that back in the 1950s so-called "aliens" were endlessly stumbled upon while they were taking "soil-samples," or how Bigfoot is so often seen crossing the road, so with the Dogmen there seems to be something stranger going on than mere, chance encounters - and we're the unwitting souls and the pawns in a bigger picture that we're not fully understanding, or perhaps not even capable of

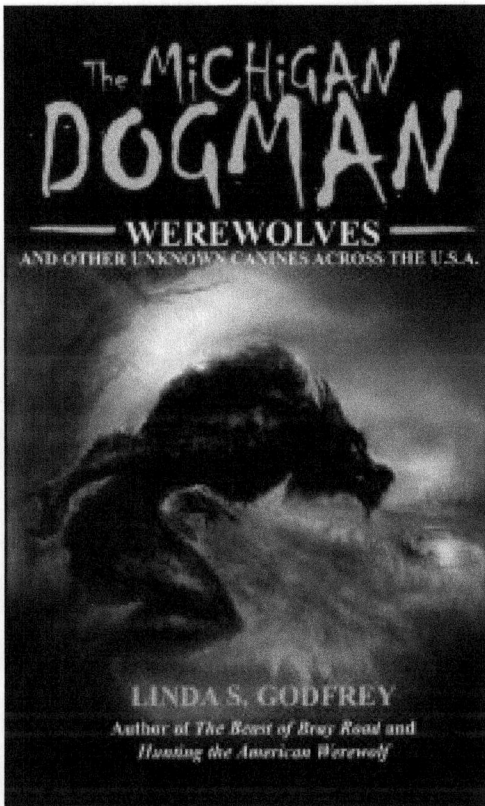

The MiCHiGAN DOGMAN
—— WEREWOLVES ——
AND OTHER UNKNOWN CANINES ACROSS THE U.S.A.

LINDA S. GODFREY
Author of *The Beast of Bray Road* and
Hunting the American Werewolf

and exit roads to highways - which is surely an upgrading of the old crossroads motif in such cases. As an aside, I'm rather reminded how, in centuries past in my home-country of England, ghostly black-dogs were seen faithfully patrolling old paths and tracks. There are other locations that play a role in such cases, too: grave-yards, ancient burial mounds, effigy sites and more. In fact, in reading Linda's Book, I was astonished to find how closely such reports mirror many, very similar, cases from the large Cannock Chase woods of England, near to where I grew up. Again, there appears to be a picture here, but we're seeing it through fogged eyes.

Then there are the attacks: you would imagine that if you were attacked by a six-to-seven-foot tall werewolf, you would sustain some pretty serious damage. After all, haven't we all gone to You Tube, once or twice, and searched "Bear attack" or "Lion attack" and seen the sheer damage that a large, powerful predator can inflict upon a dimwitted human being who thinks it's clever to try and shake hands with a 1,000 pound flesh-eating wild animal?

But, when it comes to the Dogmen, the attacks are always half-hearted - yes, there are a few torn shirts, and the very occa-sional scratch, but it's almost as if the at-tack is one of effect, one that is designed to provoke an emotional response in the wit-ness, rather than to actually turn them into a tasty dinner.

So, where does Linda stand on all this? Well, to her credit, she doesn't try and force any particular theory down the throats of her readers. Rather, she astutely recognizes that while the Dogman phenomenon is very real, it's also an issue that is infinitely diffi-

understanding, just yet. Yes, the beasts Linda describes seem real - in the way we under-stand the term - but there is distinct high-strangeness attached to such reports, too, in-cluding the locations: bridges, crossroads, and so on. Of course, any student of folklore will be aware that such locations have, for centu-ries, been associated with sightings of fantas-tic beasts (such as my own personal obses-sion, the so-called "Man-Monkey" that haunts Bridge 39 on Britain's old Shropshire Union Canal). But, the majority of the witnesses are not students of folklore - and that's an impor-tant issue.

Plus, Linda cites a number of reports where these creatures have been seen near entrance cult to resolve. So, in the pages of *The Michigan Dogman* you will get to learn a great deal about whether or not a physical, wolf-like beast could exist, in stealth, in the U.S. and elsewhere.

You will, however, also learn a great deal about Tulpas - beasts of the mind that thrive on fear and high-states of emotion; about ancient, paranormal Guardian-like creatures - the equivalents of supernatural watch-dogs, perhaps conjured up centuries ago from who knows where, and that still roam the landscape to this very day; about skinwalkers and bearwalkers; about ley-lines and the associations these beasts have with water; and even about ancient, mighty wolves that may not be as extinct as many presume them to be.

If you're already acquainted with the work of Linda (and if you aren't, why not?), then *The Michigan Dogman* is a book that you'll definitely want to read. And, if you're new to the world of real-life werewolves and this book is your first taste of what it's all about, then you're in for a real treat too! As a first-class book that offers both numerous cases of a definitively werewolf nature, as well as a variety of thought-provoking explanations to try and explain the phenomenon (or, maybe, phenomena is a better, and more accurate, term), *The Michigan Dogman* should have pride of place on the bookshelves of everyone interested in strange and unknown beasts, ancient legends, folklore and mythology. You'll find all that - and much more, too - inside its packed pages.

Linda Godfrey's *The Michigan Dogman* is published by Unexplained Research Publishing Company. Nick Redfern can be contacted at http://nickredfernsbooks.blogspot.com

Like many classic monsters, the modern view of the werewolf is far removed from legends. Most people today imagine a werewolf as a bipedal creature; a wolf-headed hairy beast that stands and runs on two feet like a man, and has hand-like claws.

In actual legends both in Native American, European and Asian law, a werewolf is a person who transforms wholly into a wolf. The werewolf is physically indistinguishable from other wolves. The idea of a 'wolf-man' first arrived with Hollywood in 1935 with the film *The Werewolf of London*, in which the creature looks more like a caveman with fangs than a legendary werewolf. This set the mould for other films like *The Wolfman* in 1941, wherein the beast looks like an anthropomorphised Yorkshire terrier. This was followed in 1942 by *The Undying Monster* with the fanged caveman look again.

Silver being lethal to werewolves is also an invention of the movies, first making its début in *The Wolfman* where Lon Chaney Jr's beast is killed with a silver-headed cane rather than a bullet. Silver bullets are now the stock in trade of werewolf killers in the movies. In legend, werewolves were as vulnerable as real wolves to ordinary weapons. However, also in legends worldwide, silver has long been thought of as having powers against evil.

Finally, the transformation under the full moon:

> Even a man who is pure in heart
> and says his prayers by night
> may become a wolf when the wolfbane blooms
> and the autumn moon is bright.

So runs the rhyme in *The Wolfman*. *The Werewolf of London*, however, introduced the idea to modern audiences. But, once again in legend, a werewolf can transform at any time of day or night. Some were said to have magical items like belts or pelts of wolf hair that were donned in order to transform, whilst others were said to be sorcerers who transformed through magic ritual. Some were even put under a curse to transform them into a monster.

Interestingly the Japanese werewolf Hito-okami is a wolf taking-on human attributes and was thought of as a positive beast and a messenger of the gods. **RF**

AQUATIC MONSTERS LOG BOOK

BY OLL LEWIS

Nesski the Killer

Lake monsters rarely live up to the term 'monster' usually turning out to be benign but very large fish and guilty of no greater crime than taking the occasional duck from beneath the waves. Rarely do they cause human injury, in the last 20 years few but the giant pike of Llangorse lake have been known to have actually caused hospitalisation, and cases where a monster has actually killed people are even rarer.

That is not to say there have not been times when a creature has been rumoured to have killed people, but often there has been nothing but local rumour and folklore to explain the disappearance of people in remote parts of the world, but these often have simple much more mundane explanations. One such example is the Congo 'river spirit' Mamawatti, investigated by extreme angler Jeremy Wade in his TV show *River Monsters*.

Wade came to the conclusion that rather than being taken by a giant fish it was more likely that the risky fishing practices employed by the Congolese tribesmen in the area were more to blame for the deaths than the giant catfish in the area. However, there are times when it looks like folklore about killer lake monsters could actually be true.

The most recent example of this occurred in Lake Chany, Siberia in July 2010.

Two elderly friends, Imamentinom Naurusovym and Vladimir Golishev were out on the lake fishing. They were familiar with the lake and, as experienced anglers, had fished the waters of Lake Chany many times before. This trip would be unlike anything they had ever experienced before and would end in tragedy for Naurusovym. Golishev tells the tale of what happened:

> "I was with my friend Imamentinom some 300 yards from the shore. He hooked something huge on his bait and he stood up in the boat to reel it in. But it pulled with such force that it overturned the boat."

In the chaos that ensued Naurusovym was pulled beneath the waters by the animal he had hooked and never surfaced. His re-

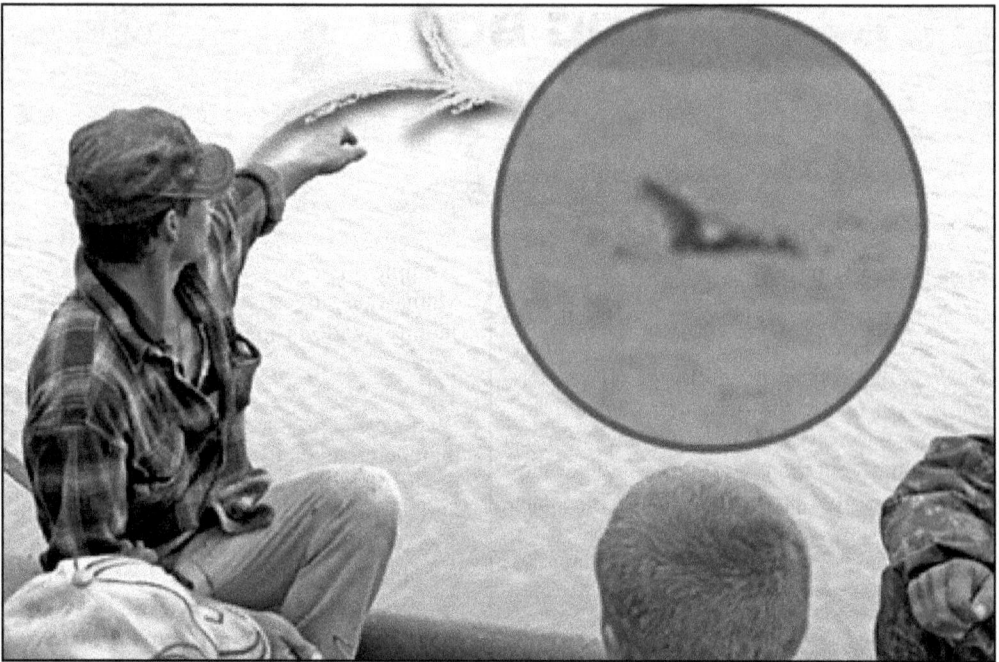

This image accompanied most of the British newspaper reports of the Lake Chany tragedy, and apparently originated with a Russian newsagency. Whether it shows the Lake Chany creature, or whether it is a fake concocted by the art department of an obscure newspaper somewhere deep inside Mother Russia is unknown.

mains have never been found leading to speculation that he was eaten by the creature. As an isolated case it might be tempting to assume that could just be a case of somebody faking their death or, worse, of a murder being blamed on a local legend in an attempt to subvert the course of justice, but

this is certainly not an isolated case.

Three years ago a 33 year old Russian sol-

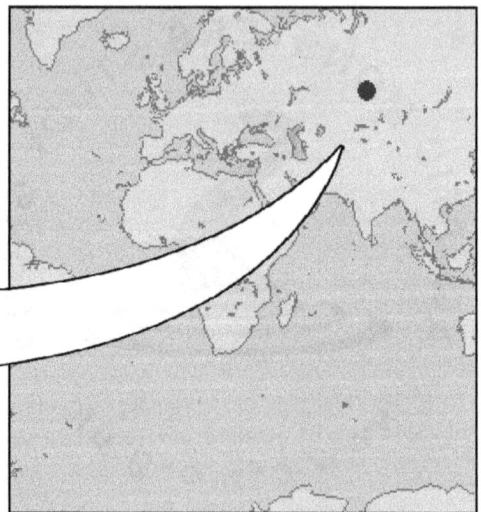

dier named Mikhail Doronin disappeared on the lake in similar circumstances and, since 2007, 19 people have drowned in the lake with bodies only turning up in a handful of cases.

Of course one cannot expect to find the body of someone who drowns in a large body of water in every case but the apparent low number of bodies recovered could well indicate that the monster of Lake Chany has figured out that humans are easy prey.

Turtle Power

In April 2011, one of the world's largest, and possibly oldest freshwater turtles was captured in Hoan Kiem Lake, Hanoi in order to administer medical treatment. There are fanciful tales that the turtle, a rare *Rafetus swinhoei*, is over 2000 years old linking it to a turtle god that helped to defend the city. Such an age would be unprecedented, but locals believe the turtle may be the same one that is mentioned in a 500 year old legend where a hero returned his sword to the lake and a turtle came up from the depths to retrieve it. Even this is unlikely as, even if the legend has a factual basis rather than a localised retelling of the King Arthur legends, up until the mid-20th Century there were several turtles present in the lake so even if that turtle survived into modern times it could have been any number of diseased turtles.

For the last 50 years or so, however, it has been generally accepted that the Hoan Kiem turtle was the last turtle living in the lake and at one time it was believed to be the last extant *Rafetus leloii* turtle, it has since been concluded that the *R. leloii* is a synonym of *R. swinhoei*, which is also a critically endangered species, and this has

given hope that the turtle may be used in a breeding campaign to preserve the species for several more generations.

Since the turtle was captured it was taken to a specially made enclosure on an island on the lake where it is currently undergoing medical treatment and observation away from the polluted silt of the lake and away from the possibility of being hit by boats, two things which are thought to have contributed to its current poor health. This is where the plot thickens as some people have since claimed sightings of another large turtle in the lake. With no photographic proof of this as yet, experts are divided as to whether there are more *R. swinhoei* in the lake of which they were previously unaware. The scientist that named the turtles *R. leloii*, Professor Ha Dinh Duc, is sure that the captured turtle is the only turtle of its kind in the lake whereas the turtle zoologist Dr Peter Pritchard is a bit more cautious suggesting that the lake could, in theory, be home to as many as five more turtles.

Come On and Join Your Fellow Man

After receiving a freedom of information request the Royal Navy admitted in May 2011 that it does not keep track of sea monsters centrally in the same way that the Ministry of Defence kept track of Unidentified Flying Object reports. The Navy's statement read:

> "The Royal Navy, and MoD, in general, do not maintain any form of central repository of information purely devoted to sea monsters. Personnel might be inclined to record unusual sightings in ship's logs, but there is, as far as we know, no actual requirement for them to do so, and it would be beyond the resource constraints of an FoI request to check every line of every RN log book for any such references since 2005.
>
> However, the RN does invite people to report sightings of marine mammals, and it's possible that this could include unusual sightings."

It is hardly an unexpected revelation, as a U.F.O. could easily turn out to be an enemy spy plane or weapon then it falls clearly within their remit, a sea monster or unknown species of aquatic animal would not unless it actually presented an obvious danger to ships. It is admirable that the Navy does encourage people to report sightings even though they are not centrally recorded.

Bownessie Gets Busy

In recent months there have been several new sightings reported of a supposed large creature in Lake Windemere. The creature, dubbed 'Bownessie' by the press, has been spotted sporadically since at least the 1940s but only came to light in the press a few years ago when photographs were published showing something creating a large wake behind it.

As fewer sightings are being reported from Loch Ness in recent years (which, as I have mentioned previously in these pages, does not necessarily mean there is less activity there just fewer reports of sightings making it as far as the papers) Bownessie reports seem to have taken its place in English newspapers.

The first reported of the recent rash of sightings was an encounter by two kayakers with the creature on the 19th of February 2011. The pair, Tom Pickles and Sarah Harrington, saw something with four

MYSTERY OF THE DEEP

NEW LOCH NESS MONSTER?
IS PHOTO REAL OR FAKE?

abcNEWS.com

humps moving across the top of the water at about 10 miles per hour. They claim to have taken a photo of the beast on a camera phone that they had in the canoe with them (a bit risky if you ask me as these things tend not to be waterproof) but it is not the most convincing of photographs and experts have said that the file size is too small to know if it has been tampered with or not.

Regardless of whether the photograph is genuine or not Pickles described what he claimed to have seen to reporters:

"I thought it was a dog, Then I realized it was much bigger and moving really quickly.

Each hump was moving in a rippling motion and it was swimming fast. I could tell it was

much bigger underneath from the huge shadow around it."

His fellow eye witness, Harrington, continued:

"Its skin was like a seal's, but its shape was abnormal -- it's not like any animal I've ever seen before. We saw it for about 20 seconds. It was petrifying. We paddled back to the shore straight away"

The next reported sighting came less than five days later and was made by 61 year old Brian Arton who was on holiday in the Lake District, with his wife, from Hovingham, North Yorkshire. At first he thought the thing in the lake, which he said looked exactly like the photograph taken by the kayakers was an inanimate object like a pontoon, log or an odd shaped buoy, until it disappeared beneath the water. Another possible encounter with Bownessie was reported that February, but happened in July 2009, was witnessed by hotel owner Thomas Noblett.

Noblett was swimming in the lake in training for a swim across the English Channel when he thinks the creature swam past him and his friend Andrew Tighe, who was following Noblett in a rowing boat.

> "We were going across the deeps, the deepest part of the lake,"

Noblett related.

> "It was really calm. Then, all of a sudden, I felt something go past the back of my legs. It felt like a cruiser had gone past.
>
> Suddenly, this wave then lifted me up. I stopped and asked 'what the hell was that? Get me out of here'."

Tighe quickly helped Noblett into the boat and recalled:

> "Looking back at that morning it is all a bit surreal. I was most concerned about Thomas. I didn't know what it was."

At first the men claim to have had no idea what could have been behind it but after Noblett's hotel's chef told them about a report of Bownessie being seen nearby Noblett swiftly came to the conclusion that he had encountered the monster.

In recent years there have been several investigations and expeditions by cryptozoologists to find evidence of the monster including expeditions by ourselves and separate unrelated expeditions by Linden Adams.

He's Not the Monster, He's a Very Naughty Bryozoan

In November 2010 the community around a man made lake near Newport, Virginia, U.S.A. was awash with rumours concerning a strange 4 foot long brown spongey mass that had been found there.

Suggestions of what it could be ranged from lake monster to a discarded alien cocoon. The man who found it, Charlie Schmuck told the story of its discovery to the press:

> "We took a stick and poked at it, and it was spongy and squishy. And that's kind of weird... The texture appears to be that of a rock with algae spots on it — it is brown and yellow, with a pattern of some type."

It was analysed by the Virginia Institute of Marine Science who were able to identify it as an outstanding example of a colony of the bryozoan *Pectinatella magnifica.*

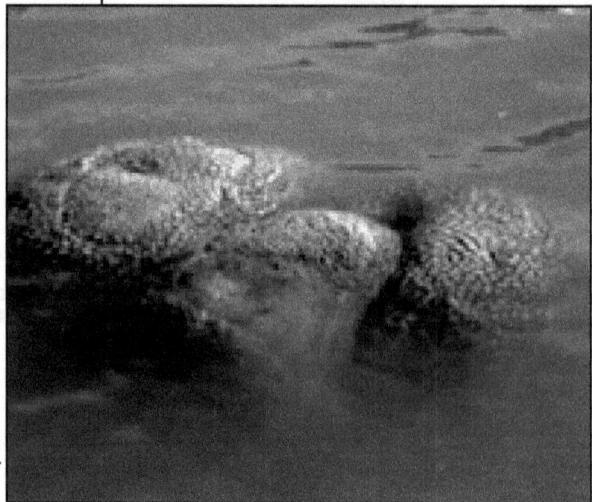

Great Snake

A giant snake has recently been spotted by several witnesses in Billings Reservoir. Grajau, Parque Residencial Lagos, Brazil. The snake is, according to the witnesses, more than five metres long and thought to be an anaconda. These sightings have, as one might expect, been causing considerable panic in the urban area that surrounds the reservoir and the local authorities have stepped in to put up signs warning locals that might use the reservoir for swimming, fishing and boating about the large and potentially dangerous snake. Some witnesses claim to have seen the snake eating mammals as large as capybara and dogs, as the president of the friends of the park association, Vera Luca Baseline, relates:

"This is a monster. It sticks his head out of the water and then, dives, lifting gigantic tail. Even a dog, she eaten."

This has led to concern that if the snake can take animals of that size there is nothing to prevent it from taking a small child swimming in the lake.

Locals disagree as to when and how the snake arrived in the reservoir, with some subscribing to a slightly sketchy sounding story about an unnamed Japanese man dumping the snake in the reservoir only a few months ago and others insisting that the snake has been in residence since at least 2003 and may have had offspring.

The Paignton Seamonster

On the 27th of July 2010 a lady named Gill Pearce claims to have spotted and photographed a sea monster just off the coast of Paignton, Devon. Pearce reported the sighting to the Marine Conservation Society believing it to have been a turtle, however the stated behaviour of the creature was unusual for such an animal as the MCS's Claire Fischer explained:

"Gill Pearce spotted the creature about 20 metres from the bay at Saltern Cove, near Goodrington.

It was observed at about 15.30 on 27 July but by the time she had got her camera it had moved further out. She spotted it following a shoal of fish which beached themselves in Saltern Cove.

The creature remained in the sea, then went out again and followed the shoal - this indicates it's not a turtle as they only eat jellyfish. We would love to know if other people have seen anything like this in the same area and can help clear up the mystery."

The pictures are quite grainy but the witness claims that the head looked reptilian and the creature appeared to have flippers, which would rule out the possibility of it being a seal or a whale, according to the MCS.

Looking at the pictures I think it looks remarkably like a sandbank or rock poking above the surface of the waves, the apparent movement out to sea reported could just be the optical illusion formed by the tide coming in more and there being nothing to get a size or distance reference to close to the 'animal'. The witness also says she saw the 'animal' stalking the fish and not chasing after them which could mean that it was staying still in the same position as if intently watching the fish waiting for an opportune moment to strike.

If not a sandbank or rock then judging from the pictures alone it would appear to have the wrong shape to be a turtle.

MYSTERY CATS DIARY

DE ROCCO FILES: FAIRFIELD COUNTY MOUNTAIN LIONS

As of March of this year, the eastern mountain lion (*Felis concolor couguar*) was officially declared extinct by the U.S. Fish and Wildlife Service. After years of an elusive presence on the endangered animals list, the eastern cougar was finally branded with the status of extinct.

But was this ruling correct? Recent mountain lion sightings in southern Connecticut suggest that the call may have been premature. Beginning on June 5th, credible sightings began to filter in of a mountain lion roaming the town of Greenwich, CT. Officials from the Department of Environmental Protection (DEP) were first alerted by three credible sources at the Brunswick School in Greenwich that had reported seeing the mountain lion. The sources consisted of two teachers at the school and a passing EMT. Upon further investigation by the Greenwich police department and the DEP, it was verified through paw prints that the creature was in fact a mountain lion. The lack of biological evidence made it impossible to determine exactly what species of cougar the animal was, though, drawing an investigation rapidly to a close. I spoke to the Director of Conservation of the Town of Greenwich and she claimed that the DEP was finishing up an investigation but had told her that the trail appeared to be going cold.

THE CENTRE FOR FORTEAN ZOOLOGY

MYSTERY CATS STUDY GROUP

That is, until more sightings of the lion began to crop up throughout the town. And then, in a surprising turn of events straight out of a riveting novel, an unnamed driver on Route 15 collided with a feline of some sort, killing the creature near Milford, CT on June 11. DEP officials arrived on the scene and officially verified the dead animal as a mountain lion before sweeping it off to a secure facility for further testing. They released information that stated that the cat was a male, less than six years old, and with no evidence of domestication. The distinct lack of neutering and the extremely lean, muscular body suggested that the creature had spent a long time in the wild.

Regardless, the DEP has stated that they believe that it had been released from captivity accidentally and the Environmental Conservation Police (EnCon) are currently following a "lead" in New York. No one in any state has offered any intelligence on where this creature would have been released from though. The EnCon's current story is that they suspect a private owner in New York may have inadvertently released the cougar a while back, and it stayed in the wild, cross-

ing the New York-Connecticut border and finally being spotted in Greenwich before its death in Milford.

This depiction of events rings a little hollow, though. The lack of evidence to suggest any previous domestication of the animal is but one part that does not quite add up. A DNA analysis of the creature could easily verify whether or not it was in fact a cougar imported from Florida, as suspected by the DEP, or a true remnant of a dying population of eastern mountain lions. Unfortunately, the DEP refuses to allow any access to biological samples secured from the animal to an outside source. After a series of increasingly frustrating attempts to contact the Connecticut Department of Environmental Protection about the mountain lion, an associate of mine was kind enough

to walk into the Hartford headquarters and ask point-blank about securing DNA samples.

He was told that no one was allowed samples. The DEP and EnCon want to handle the entire investigation and plan to release a statement about the mountain lion once they have enough evidence to paint a believable picture.

My associate left the office disheartened and phoned me to deliver the news. The refusal to allow any independent verification of their findings only further increased my skepticism about the DEP's story.

It is true that to keep the cougar on the endangered list costs the state more than to declare it extinct, but this does not seem like enough of a rationale to intentionally control the analysis of this animal.

Do they fear that a third-party investigation would reveal it to be an eastern cougar that persisted? It is hard to imagine that this could be the case, but I continue to maintain a measure of healthy skepticism about their story.

Interestingly enough, control of the mountain lion may be slowly slipping from the DEP. Sightings have once again erupted in Greenwich and in nearby Fairfield. The most recent sighting took place in Fairfield on the morning of the 16th.

A large feline was seen crossing a residential road at 8 AM. Other sightings include a report from an NBC Connecticut crew that apparently had to slam on the brakes to avoid striking the lion on Route 63.

The Department of Environmental Protec-

tion has stated that it unlikely that these sightings are valid, which is quite probably true. People see what they want to, so mountain lion hysteria is not an unlikely cause for a majority of these sightings.

But the overwhelming amount of reliable witnesses for many of these sightings has forced even the DEP to send investigators to search for scat and prints in Greenwich. The current statement of the DEP is that there was only the one cougar, and the rest of the sightings have been a "false alarm."

A press release on the Milford lion's DNA analysis is forthcoming. The investigators also plan to conduct a necropsy to look for any evidence of domestication. They claim that they will release their findings in about a week.

My personal opinion is that their story will be that it was a Florida panther imported by an unnamed collector in New York that escaped and found its way into Greenwich before being killed in Milford.

I doubt that there will be an opportunity for the CFZ to corroborate these findings because I do not foresee the DEP allowing access to DNA samples in the immediate or distant future. My inquiries into their investigation have proved stifling.

The hope remains that the recent sightings are not the products of hysteria but instead are sightings of the mate of the dead lion.

My investigation into this topic is far from complete and I plan to pursue any avenues to get closer to these mysterious feline entities that have started to haunt Fairfield County. **WILL DE ROCCO**

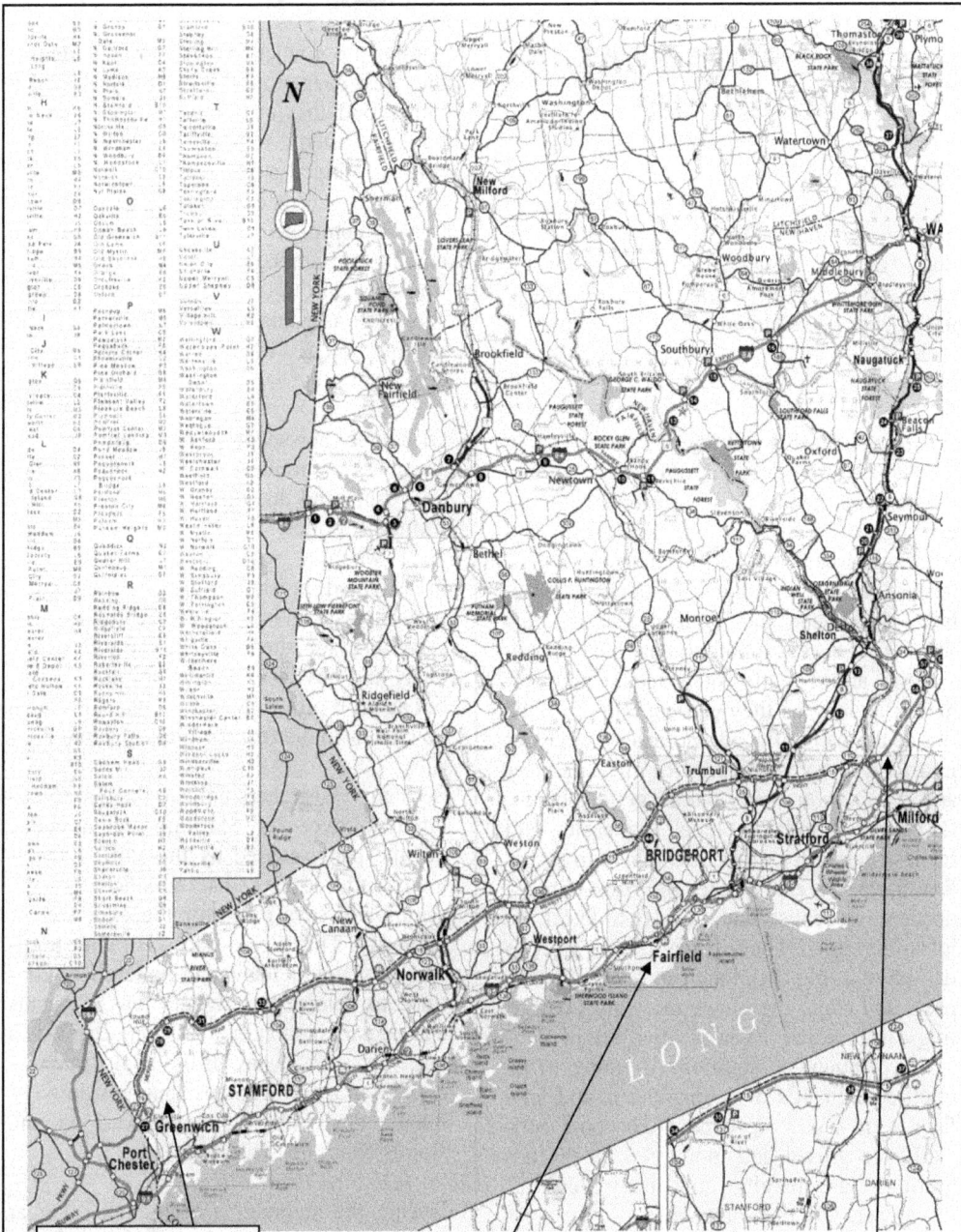

Greenwich, CT
Hub of original and
recent sightings

Fairfield, CT
Site of some recent
sightings

Milford, CT (Route 15)
Site of death of Milford
Lion

BRITISH BIG CAT ROUND UP

'Woman's panther terror', 'Ex-policeman in big cat sighting', 'Oh my god, that's not a dog', 'Best ever big cat sighting'; the headlines, even after thirty years of press interest, have never ceased to amaze.

Sadly.

It's no wonder that sightings of exotic cats in the UK are still relegated to folklore. Despite the fact that the last year has produced a 'big cat' sighting from every English county, no sceptic has been swayed in their non-belief, despite the usual collated evidence of scat, hair, sheep/deer and even dog kills, plus some half-decent photographs. Of course, sceptics aren't really interested in any evidence, as long as they can continue to poke fun at the camouflaged anoraks who sit out in a field in-between episodes of *Eastenders* and *Coronation Street* before getting bored. Of course, in today's climate it's easier, and far less hassle for most researchers to set up a trail camera and bugger off back home – do they not realise that cats often detect the infra-red light ? Then if they do get a photo or film of a 'big cat', they can rush off to the press and claim they have the Holy Grail of 'big cat' research. Even so, no piece of footage or amount of sightings is going to be good enough. It's a vicious circle, and let's face it, if such 'big cats' do become officially recognised and monitored by the authorities, just imagine how many 'big cat' researchers will be redundant!

When various news sources commented in January 2011 that an ex-policeman had seen a 'big cat' in west Wales, this sighting - according to 'council chiefs' - was hailed as the "best ever proof." I've always found it rather odd that a member of the general public is often dismissed should they see a 'big cat', but if a policeman or other such like, see a cat then they must be credible - this is complete rubbish of course. The amount of sightings each year of so-called 'big cats' in the UK clearly suggests large cats such as puma, black leopard and lynx inhabit the UK - it's no mystery. It's no big deal anymore.

Of course, newspapers, and even some researchers continue to get their species mixed up, and this inadequacy - especially from the police - was confirmed in May 2011 when a 'white tiger' sighting was investigated in Hampshire. Not only did police waste a lot of resources and time searching for the creature, which hilariously and obviously turned out to be a toy, but again it proved that the authorities have no clue as to the British 'big cat' situation.

So many researchers believe there are government cover-ups regarding such animals, but the panic caused by the stuffed specimen, and the fact that a helicopter was scrambled, says a lot for the expertise of authority. This type of story is rather tragically not unique, as some researchers continue to believe that tigers, lions, cheetahs and jaguars inhabit the UK. On occasions a large cat such as any of the above may escape from a zoo, but these animals do NOT inhabit the wilds of Britain. And yet, the stories continue to crop up.

During the summer, a woman jogging at 7:30 am through a rural area of Maidstone, Kent, saw a large cat, of which she managed to get one picture on her mobile phone. Again, sceptics argue as to why there aren't more photos of large, exotic cats without realising the shy nature of such animals, and the fact that they tend not to pose for photographs! The photo shows an animal that is approximately four-feet long. This photo was sent to four separate experts - Richard Freeman, Karl Shuker, Darren Naish, and Jonathan McGowan. Darren and John both agreed that the photo shows a puma, whereas Karl Shuker felt the animal was too "delicate" in appearance and stated that something wasn't right with the photo, but commented no further.

Richard Freeman commented that the animal may have been a Cornish Rex cat (see insert) although they do not reach four-feet.

I believe the animal is a puma, especially as

there had been several other reports around the same time.

Meanwhile, during the early part of the year a Maidstone woman named Sharon Ramsden set up a trigger camera in her back garden and one night it snapped a photo of a large, black cat.

When Sharon submitted the photo to several researchers and websites she was ridiculed despite the fact that no-one investigated the report. The photo shows an animal leaning on its haunches, peering in the direction of Sharon's garden corner where she keeps her pet bird. Compared to the tulip fencing, which is three-feet high, the animal is three-feet in length (in a sitting posture) and almost the same height. Opin-

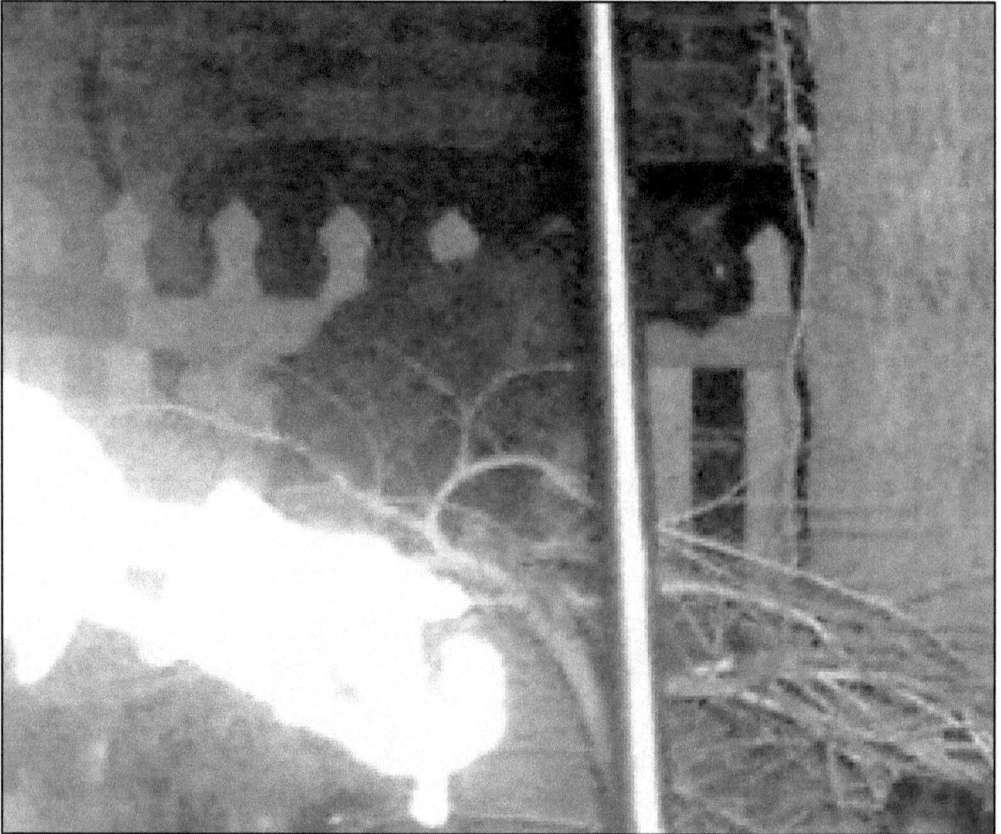

ion has been divided as to whether the animal is a large feral cat or a sub-adult leopard. In the same area scat has been found, as well as a scratch post. Interestingly, scat found at Blue Bell Hill, near Maidstone, was commented on by a biologist as possibly "being an owl pellet", but when the sample was sent for analysis to a local university it was returned as "leopard faeces". (Opposite, Top).

On several occasions throughout 2010 and 2011 there were reports of foxes found eaten along the line of the River Medway. Some were stripped clean, others simply displayed a fierce bite to the throat. What had killed them? (Opposite below).

On two occasions horses were reported as been attacked by a large cat which left marks to the throat of the victim (which on both occasions fortunately survived) and scratch marks on the hindquarters. (See po.50).

During the first two weeks of June 2011, I received twenty-four reports of differing cats from across the county. These were mainly from the areas of Ashford, Dover and Maidstone, but there were also more from Sussex (one sighting concerning a lynx was filed by the police, and there was

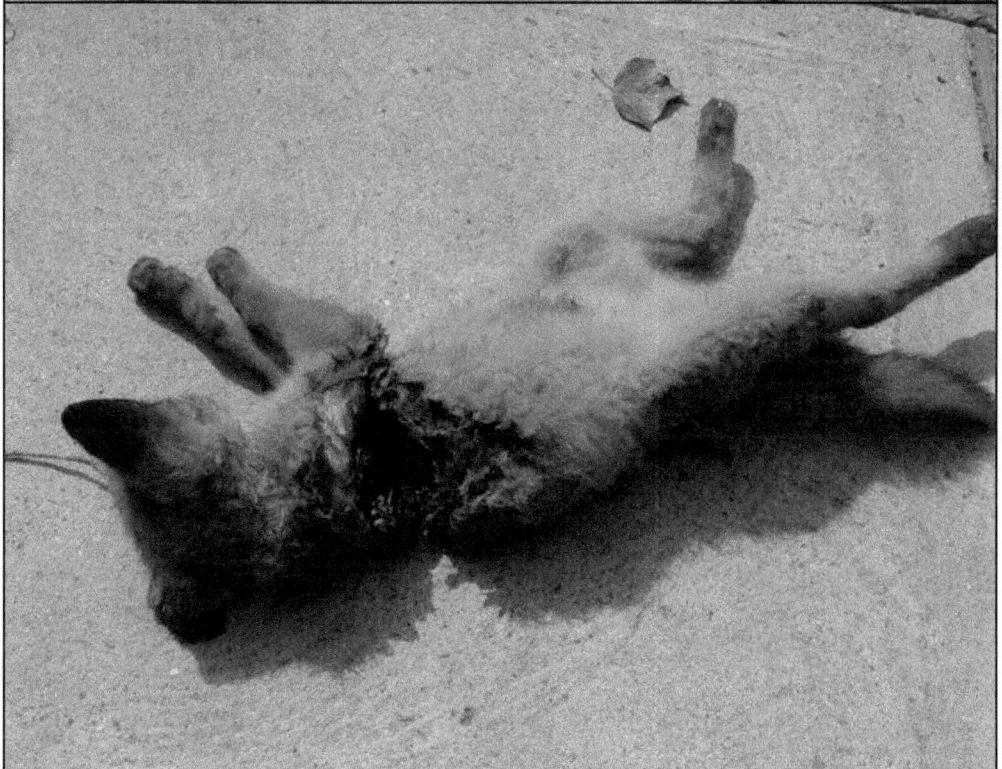

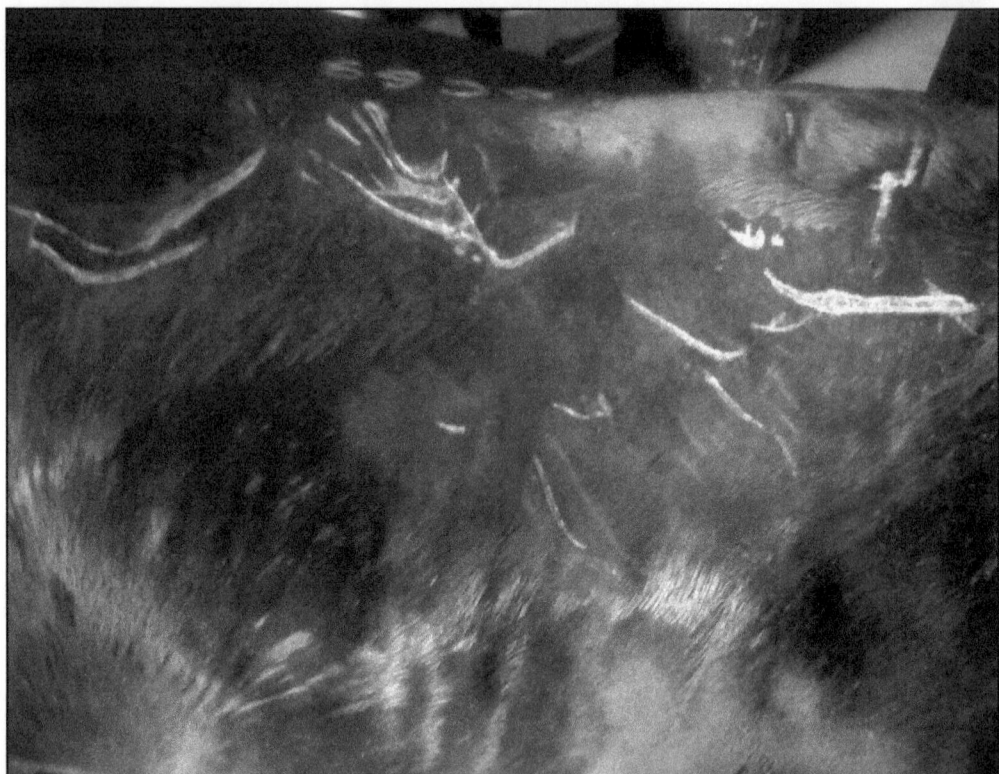

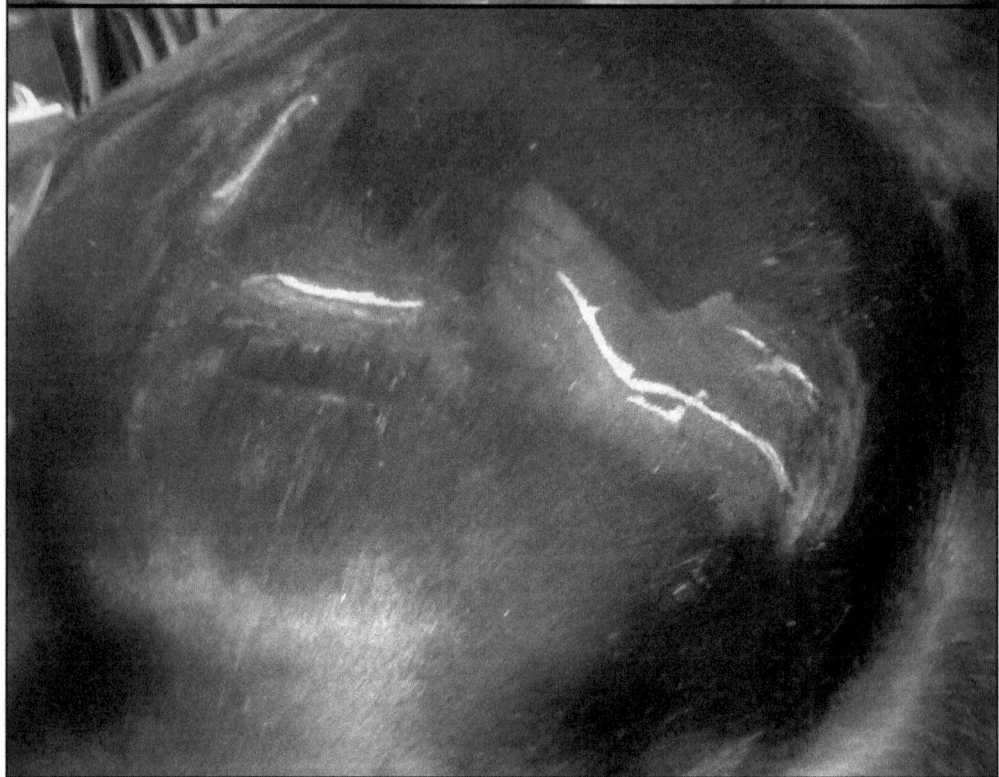

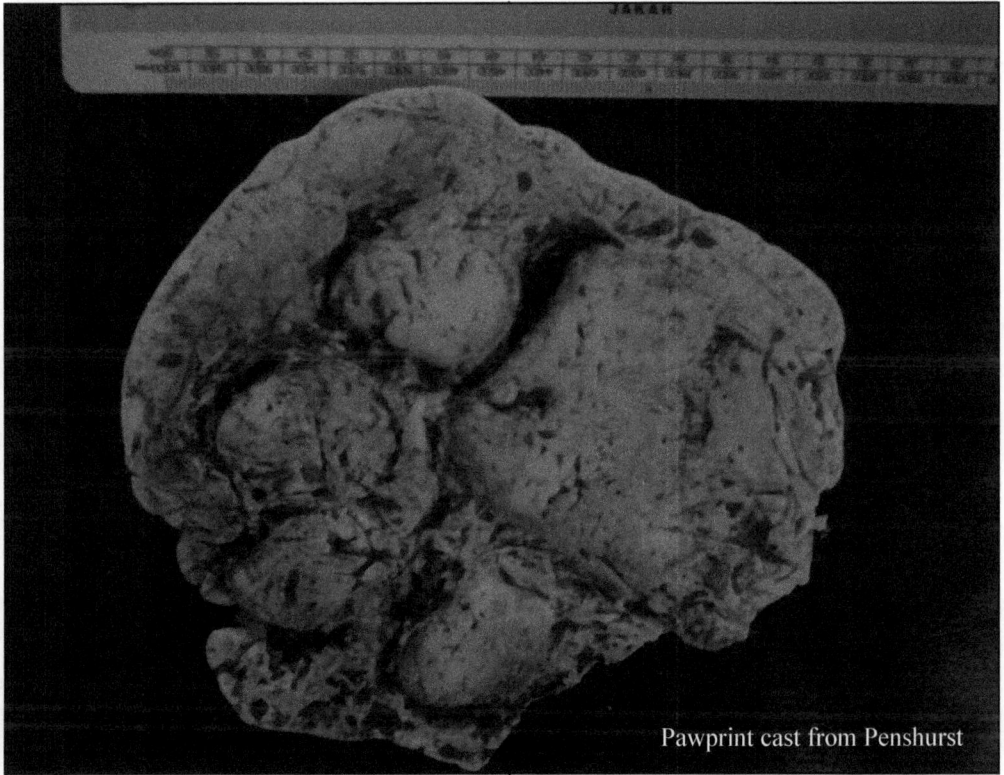

Pawprint cast from Penshurst

another report of two deer that were found up in a tree after sightings of a black leopard). Reports also came from the London outskirts and Surrey (one concerning a female jogger who, whilst running through woods at Dorking, heard a large animal leap from a tree with a thud).

Paw prints were found at Tunbridge Wells and casts were made of these. During the same month two black leopards were seen together in the Forest of Dean in Gloucestershire, and further sightings came from Northamptonshire, Yorkshire, Shropshire, Wiltshire, and Scotland, etc, etc. But again, such reports, many of which made the local newspapers, were again plagued by some hilarious headlines and comments from researchers, such as "there aren't many cats out there....just two or three...no more than half a dozen though", and "sightings have increased in the last ten years", and of course the classic, "there are no big cats in the wild but there could be an escaped animal", which was said by a Kent zoologist of all people - that's certainly one way to start a panic!

Whether people believe in 'big cat' sightings or not, everyone loves a 'beast of...' story, so much so that I'm currently writing a book on 'big cat' headlines from around Britain.

For me, the most important aspect with regard to investigating such reports is consistency. There have been too many far-fetched reports and vague reports taken seriously over the years.

There are still those who believe that such cats are paranormal, or that leopards and puma are breeding, or that tigers roam the suburbs, but if we look at the consistency (or lack thereof) of these reports then they must be rubbished. In the south-east of England there have been no consistent reports to suggest leopards of normal pelage, no consistent reports to suggest hybrid cats, i.e. puma x leopard, and no consistent reports to suggest there are 'black' pumas.

Even the darkest pumas do not appear jet-black and would have a lighter underside.

So, where do we go from here?

The answer, sadly, is nowhere. If a child is attacked, only that particular case will be investigated. Such animals will never be tagged.

They will never be 'officially' monitored, and never be fully taken seriously. This was a concern raised by the late Quentin Rose - a professional wild animal trapper whom the police and Ministry of Defence often consulted; he also helped me with my research.

He estimated that there were, during the late '90s, approximately 100 large cats across the British Isles.

Unlike most researchers, he looked at the negative and very serious aspect to the situation, rather than the mystery. Sadly, mystery is all it will ever be, because the British 'big cat' situation is stale, and only revived in silly headlines. Where do we go from here, because the evidence - however positive -

isn't good enough anymore?

Whilst today's breeding populations of lynx, puma, black leopard, and some smaller cats are certainly the product of those released in the '60s and '70s, I'm amazed at how researchers still can't fathom out as to why such animals roam Britain.

Yes, there have been sightings of large cats dating back centuries - these would have been escapees from menageries, zoo parks, circuses etc., but why do researchers always look for a more esoteric solution? It is simply because they are constantly at odds with their own theories and struggling to find a consistency.

Researchers often state that it's a mystery that black leopards roam the UK – it isn't. I have on record cases from Scotland, Bedfordshire, Devon, Kent, Sussex, Yorkshire, London, Surrey, Norfolk and a few others, of black leopards being kept as pets from the 1930s onwards.

I also have on record several cases of such animals - most as cubs being released: one from the 1940s concerned a London family who let their new 'pet' go because it ate the neighbour's domestic cat.

Black leopards were being sold at auctions in the Victorian era and any black leopards purchased from *Harrods* would not have been publicised - this was proven when I visited *Harrods'* archive department, although *Harrods* are far from being responsible for the animals we see today.

More information on this can be found in my upcoming book *Mystery Animals of the British Isles: London.* **NEIL ARNOLD**

watcher of the skies

CORINNA DOWNES

Bernard Heuvelmens said that "*cryptozoology is the study of unexpected animals*". We are adding this section as a new feature to *Animals & Men* and in this issue are some of the rarer birds that have been seen in the UK since the last issue of the journal was published. Over the coming issues we hope to be able to list birds that are very rare and also some that may be particularly unexpected.

The British list comprises all those bird species which have occurred in a wild state in Great Britain, and in general the avifauna of Britain is, of course, similar to that of Europe, although with fewer breeding species.

There are 587 species of birds on the British list as of 22 July 2009, with the latest additions being the Pacific Diver (*Gavia pacifica*), Yellow-nosed Albatross (*Thalassarche chlororhynchos*), Glaucous-winged Gull (*Larus glaucescens*) and Asian Brown Flycatcher (*Muscicapa dauurica*).

Glossary

British Ornithologist Union (BOU) list Category A: a species that has been recorded in an apparently natural state at least once since 1 January 1950

AEWA - Agreement on the Conservation of African-Eurasian Migratory Waterbirds

Accipitridae

February 2011 – Slapton in Devon, Ulverston in Cumbria, Hewas Water in Cornwall, Tory Island in Co. Donegal

Black Kite (*Milvus migrans*)

BOU/IRBC Category: **A**

RBA Status: **Rare vagrant**

On 1st February a black kite was seen flying over Slapton in Devon during the afternoon, and possibly one of the same was seen briefly near Ulverston in Cumbria. There was another sighting of a black kite flying west over Hewas Water in Cornwall and on 8th February one was sighted on

Tory Island, Co Donegal.

This bird is, in fact, classed as a rare vagrant, the last record in Britain and Ireland being between 5th and 14th September 2007 at South Slob, Co. Wexford. The bird breeds in most of Europe, Asia, Africa and Australia and winters in south Eurasia to southern Africa and Australia. Its natural habitat is open wood, savannah, steppe and human habitation, where it scavenges on scraps. It was first recorded in Northumberland in 1866.

Anatidae

March 2011 – Bamburgh, Northumberland
Black Scoter (*Melanitta americana*)
BOU/IRBC Category: **A**
RBA Status: **Extremely rare vagrant**
A black scoter was discovered on 14th April on the sea off Bamburgh, Northumberland.

This is only the second record for England since one was picked up at Leighton Moss in Lancashire on 16th May 2007 and then released at Jenny Brown's Point. It is classed as an extremely rare vagrant to our shores. It lives on tundra lakes and on migration rivers, lakes and coasts and breeds in Canada, Alaska and north eastern Siberia. It winters south to the United States and feeds mostly on molluscs, but also takes insects, and fish eggs when in freshwater habitats. The first record was in Lothian in 1987. This bird is known to occasionally perform an odd display while swimming where it flaps its wings with its body held up and punctuates this with a downward thrust of head as if its neck is momentarily broken.

March 2011 – Sandy Water Park, Carmarthenshire
Blue-winged Teal (*Anas discors*)
BOU/IRBC Category: **A**
RBA Status: **Rare vagrant**
From the 20th to 22nd March a drake blue-winged teal was seen at Sandy Water Park in Carmarthenshire which makes it the second for that county following one in March 2000. The bird is classed as a rare vagrant, the most recent record in Britain and Ireland being in 2009 between 23rd September and 26th October 2009 at Haverton Hole in Cleveland. The first record for Britain and

Ireland was made in 1860 in Cheshire. The blue winged teal winters from southern California to western and southern Texas, from the Gulf Coast to the Atlantic Coast and south to Central and South America. It can often be found wintering as far south as Brazil and central Chile. During migration, some birds may fly long distances over the open ocean, and they can become an occasional vagrant to Europe. Their yellow legs make them distinct from other small ducks.

March 2011 – Unst and Linga, Shetland
King Eider (*Somateria spectabilis*)
BOU/IRBC Category: **A**
RBA Status: **Rare vagrant**
On the 21st March one, or possibly two, drake King eiders were found on Shetland – one off Unst and the other at Linga which is between Unst and Yell. This bird is a rare vagrant to our shores and the most recent record was on 21st December 2009 at Haroldswick, Unst. It was first recorded in 1832. The King eider breeds in arctic Eurasia and North America and winters south to north Europe and central United States. It lives on tundra ponds and lakes. The female King eider does not feed very often during the 22-24 day incubation period. One female was not observed to leave her nest for seven days, until being flushed by an Arctic fox.

June 2011 – between Blackdog and Murcar in Aberdeenshire
White-winged Scoter (*Melanitta deglandi*)
RBA Status: **Mega vagrant**
The highlight of the 12th June was the identification of Britain's first American white-winged scoter between Blackdog and Murcar, in Aberdeenshire. It was present offshore with a large scoter flock - including three surf scoters. It is classed as a mega vagrant to our shores for the obvious reason above, and breeds in the far north of Asia and North America. It winters in temperate zones: on the Great Lakes, the coasts of northern USA and the southern coasts of Canada and Asia as far south as China.

Certhiidae
March 2011 – Landguard, Suffolk
Short-toed Treecreeper (*Certhia brachydactyla*)
BOU/IRBC Category: **A**
RBA Status: **Extremely rare vagrant**
A highlight for the end of the month was the discovery of a short-toed treecreeper at Landguard, Suffolk. This is an extremely rare vagrant, the most recent being on 8th

May 2005 at Dungeness in Kent, and there were only 25 recorded between 1950 and 2009. In fact, the first record was in 1969, again in Kent. It is also the first for the county of Suffolk. It was apparent that the bird was very low on fat reserves and had recently completed a long flight. Its usual habitat is woodland and town where it eats insects and spiders obtained from crevices in tree trunks that it feeds from as it moves up the tree. It breeds in the Channel Islands and France and is common from Germany and Poland southwards.

Cerylidae
February 2011 – Claudy, Co. Londonderry
Belted Kingfisher (*Megaceryle alcyon*)
BOU/IRBC Category: **A**
RBA Status: **Extremely rare vagrant**
On 6th February a male belted kingfisher was seen fishing in the river at Claudy in Co. Londonderry. One observer only saw this, but if it were to be accepted, it would be the second record for Northern Ireland, after the one shot at Dundrum Bay in Co. Down in October 1980, and which now resides in the Ulster Museum. There are three records from counties Mayo, Clare and Tipperary in the Republic of Ireland, but none since 1985. Since that year, the only other one seen was a roaming bird in

Staffordshire, East Yorkshire and Aberdeen-shire in 2005. The first ever recorded was in Cornwall in 1908. Hence, this bird is classed as an extremely rare vagrant. It breeds in North America and winters down to northern South America. As in our own kingfisher, the belted variety nests in a burrow it has excavated in a dirt riverbank. This can be up to 8 feet long or those of you who work in modern measurements – 2.5 metres long.

Charadriidae
June 2011 – Dornoch Point, Highland
Greater Sand Plover (*Charadrius lesche-naultii*)
BOU/IRBC Category: **A**
RBA Status: **Mega rare vagrant**
A summer plumaged greater sand plover was present for a second day at Dornoch Point, Highland on the 16th June, and is the 15th for our shores. It is classed as a mega vagrant, and is seen as an accidental visitor. It was first recorded here in 1978 in West Sussex and between 1950 and 2007 there were 14 records. Its breeding grounds are from the Middle East to central Asia, and it winters from southern Asia to Australia and southern Africa where it lives on stone flats, sandy riverbanks and on migration estuaries. It feeds on insects, especially beetles.

Emberizidae
May 2011 – Bolton Abbey, Yorkshire
Rock Bunting (*Emberiza cia*)
BOU/IRBC Category: **A**
RBA Status: **Extremely rare**
A male rock bunting was photographed on 11th May at Bolton Abbey, Yorkshire and if confirmed this will be the sixth for Britain. There have only been 5 records to 2009, the first being in 1902 in West Sussex. The most recent record before this one was on 1st June 1967 in Bardsey, Gwynedd. It lives in southern Europe and central Asia on open

scrub, where its diet consists of seeds – especially grasses – and in the summer it adds invertebrates to its meals.

Laridae
May 2011 – Minsmere RSPB, Suffolk
Audouin's Gull (*Ichthyaetus audouinii*)
BOU/IRBC Category: **A**
RBA Status: **Rare vagrant**
The sighting of this Audouin's gull was the first record for Suffolk and only the sixth for Britain, the most recent being between 15th and 23rd August 2008 at Huttoft Bank in Lincolnshire. It lives on the islands in the Mediterranean along the seacoasts, and eats mostly fish but also invertebrates, small birds and plant material.

January 2011 – Rainham Marshes RSPB, London
Slaty-backed Gull (*Larus schistisagus*)
RBA Status: **Mega rare vagrant**
On the 14th January the sighting of a potential slaty-backed gull was recorded at Rainham Marshes RSPB, London. If this truly was such a bird, then it would be a first for Britain. It is a large white-headed gull that

breeds on the western coast of Alaska and like other gulls is a great wanderer and travels widely during non-breeding seasons. However, it is very similar to other gull species, which can make identification difficult.

Rallidae
April 2011 – Arundel, West Sussex
Little Crake (*Porzana parva*)
BOU/IRBC Category: **A**
RBA Status: **Rare vagrant**
A little crake was discovered on 9th April at Arundel in West Sussex. The most recent record for a little crake was in 2008, between the 9th and 23rd April on Exminster Marshes in Devon, and it is classed as a rare vagrant to our shores. It breeds in southern Europe and central Asia and winters south to north

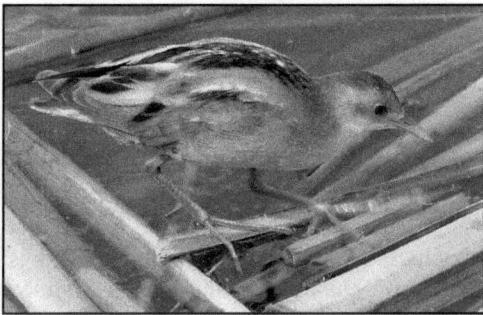

Africa and north India. Between the 1950 and 2007 range of records there were 39 acceptable sightings and it was first recorded in 1791 in East Sussex. Its usual habitat is in marsh and reed beds where it feeds on small invertebrates, and the seeds of aquatic plants.

Scolopacidae
May 2011 - Shetland
Great Snipe (*Gallinago media*)
BOU/IRBC Category: **A**
RBA Status: **Extremely rare vagrant**

The 2nd May brought the discovery of a great

snipe in Shetland. This is an extremely rare vagrant and was first recorded in Kent c. 1780. Between the record years of 1950 and 2007 there were 148 confirmed sightings, and the latest since then was in 2008 between 17th and 18th September on Speeton Moor, Yorkshire. The great snipe's normal habitat is wet grassland, marsh and damp scrub where it feeds on earthworms, molluscs and seeds. It winters in eastern and southern Africa, and breeds in north eastern Europe and northern and central Asia.

Strigidae
June 2011 - Bryher, Isles of Scilly
Scops owl (*Otus scops*)
BOU/IRBC Category: **A**
RBA Status: **Mega vagrant**
On the 2nd June came the report of a Scops Owl on Bryher, Isles of Scilly. This is a mega vagrant to our shores and breeds in southern Europe, north-western Africa, and central Asia, and winters from southern Europe to central India and southern Af-

rica. Its natural habitat is open woodland, savannah and scrub where it feeds mostly on insects and other invertebrates.

It hunts mainly at night from its perch. It was first recorded here in 1805 in West Yorkshire and between 1950 and 2007 there were 31 records.

Turdidae

January 2011 – Leigh, Manchester
Dusky Thrush (*Turdus eunomus*)
BOU/IRBC Category: **A**
RBA Status: **Extremely rare vagrant**

During January there was a very late report of a dusky thrush having been seen in Leigh, Manchester for an hour on the 8th December.

This bird is an extremely rare vagrant and was last recorded in Skomer, Pembrokeshire between 3rd and 5th December 1987.

It is a strongly migratory species and winters south to southeast Asia, mainly in China and neighbouring countries. It breeds eastwards from central Siberia.

Upupidae

March 2011 – Dorset, Devon and Cornwall
Hoopoe (*Upupa epops)*
BOU/IRBC Category: **A**
RBA Status: **Scarce visitor**

Throughout the month in Dorset, Devon and Cornwall, came sightings of hoopoes. First recorded circa 1600 the hoopoe is classed as a scarce visitor to our shores. The bird is widespread across Europe, Asia and North Africa, Sub-Saharan Africa and Madagascar. Most of the European and north Asian birds migrate to the tropics in the winter, whereas those from the African populations are sedentary all year round.

They have been known to breed north of their European range, and in the past in southern England during warm, dry summers which provide them with plenty of grasshoppers and similar insects. However , since the 1980s the north European populations have been reported to be in the decline, perhaps because of the changes in climate. Interestingly, the hoopoe's diet includes many species that are regarded as pests by humans and the bird is afforded protection in many countries because of this.

They have generally made an impact over much of their range; for example they were considered sacred in Ancient Egypt, thought of as thieves in much of Europe, as harbingers of war in Scandinavia and are strongly connected with death and the underworld in Estonian traditions. In the Bible, Leviticus 11:13-9, they are listed among those animals that are detestable and are not to be eaten.

SOMETHING LOST BE-HIND THE RANGES

RICHARD FREEMAN

The expeditionary team of Dr Chris Clark, Adam Davies, Dave Archer and myself, who had previously searched for the Russian almasty (a relic hominid) and the puzzling Sumatran orang-pendek (mystery ape or hominid), were getting our heads together in planning where to go in 2010.

Several years before, Adam had been in Tibet on the track of the yeti. Ian Redmond, Tropical Field Biologist and Conservationist, mentioned to him that there were numerous reports of the yeti in the northern Indian state of Meghalaya. Upon returning to England, Adam investigated more closely and found that a local documentary film-maker and journalist, Dipu Marak had been on the trail of the creature for some years.

I, too, had heard of the Indian yeti, or as it is locally known *'mandeburung'* - the forest man. In June 2008 BBC journalist Alistair Lawson visited the area to investigate sightings of the creature. He was impressed by the remote, undisturbed landscape and wrote…

> "If ever there was terrain where a peace-loving yeti could live its life undisturbed by human interference, then this has surely got to be it.
>
> Perhaps the most famous reported sighting was in April 2002, when forestry officer James Marak was among a team of 14 officials carrying out a census of tigers in Balpakram when they saw what they thought was a yeti."

Dipu had given the BBC some hairs he had found at a remote area called Balpakram. Upon analysis, these proved to be from a species of Asian wild goat called the goral (*Nemorhaedus gora)*. This, however, did not negate the eyewitness reports.

We decided that the CFZ team should investigate and began to lay plans for a trip to India. Adam, who is a great organiser, contacted Dipu who in turn planned guides, lodges and contact with eyewitnesses.

Jonathan McGowan was to join the four of us on this trip. Jon is an excellent field naturalist and taxidermist, as well as being the curator of the Bournemouth Natural History Museum.

On Hallowe'en 2010, we flew out to India. During the long journey, we had to call for a doctor when Chris collapsed, but after he was given oxygen, he quickly recovered. The verdict was that he has been suffering

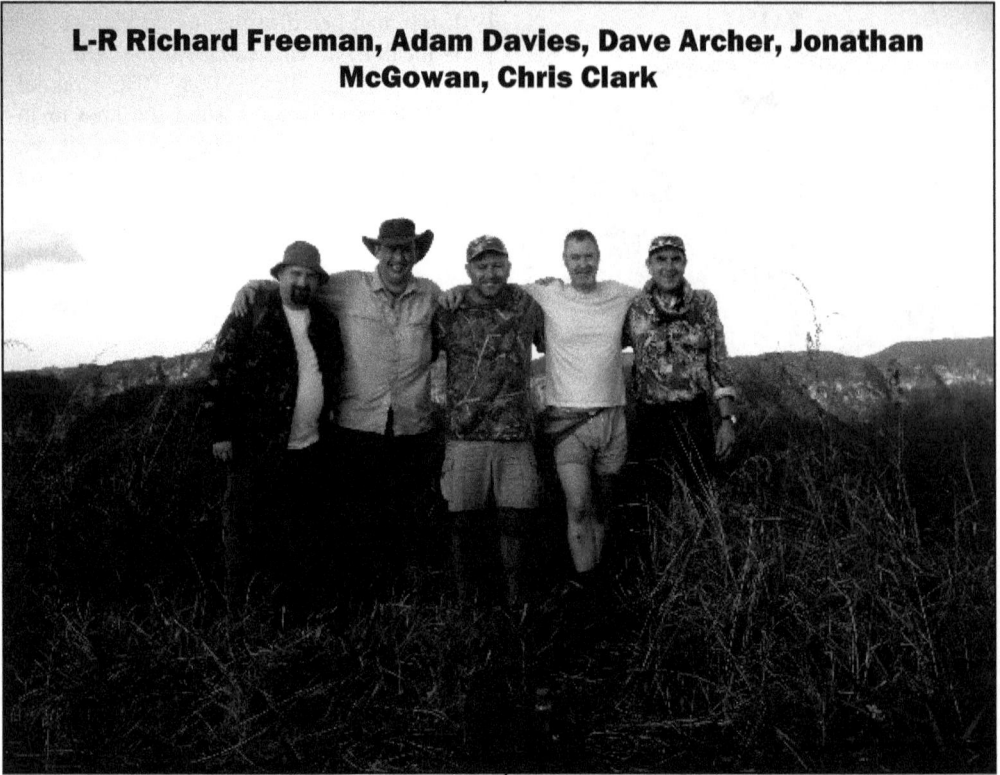

L-R Richard Freeman, Adam Davies, Dave Archer, Jonathan McGowan, Chris Clark

from a lack of oxygen on the long stuffy flight.

We arrived in the mad cacophony that is Delhi in the evening and checked into our hotel. We had an evening to kill so we arranged for a taxi to show us some of the sights of the city. Unfortunately, the taxi driver just dumped us at a western style mall presumably thinking that, as westerners, it would be the place in which we were most interested!

The following day we flew out from the surprisingly clean and efficient Delhi Airport to Guwahati in Assam. We were met at Guwahati by our chief guide Rudy Sangma, along with our assistant guide Pintu, and our driv-

ers. We then began the long journey to the town of Tura in the West Garo Hills.

Meghalaya is a mountainous state in the northeast of India. It was carved out of the state of Assam in 1972 to accommodate the Khasi, Garo, and Jaintia tribes who at one time each had their own kingdoms. The three territories had come under British administration in the early 1800s and were assimilated into Assam in 1835. Once fierce head-hunters, the Garos were among the first Indians to be converted to Christianity by British missionaries. After conversion, the tribes were largely left alone, allowing a lot of their culture to remain intact.

As the winding roads rose upwards, giving way to rocky tracks Rudy told me of some of the other strange creatures from the folklore of the Garo Hills. One creature that looms large in the Garos is the *sankuni*. This is a monstrous snake that bears a crest upon its head much like rooster's comb. Sound's familiar? It should do - the description of the sankuni matches up very well with that of the naga (the vast crested serpent I searched for in Thailand back in 2000) and the ninki-nanka (the serpent dragon of the Gambia I hunted in 2006). All are said to bear crests, be of huge size, have shining black scales, live in lakes or rivers as well as subterranean burrows, and to have an association with rain. This uncanny dovetailing of these stories made me seriously wonder if the sankuni and other monster snakes are based on encounters with a real-life species of gigantic snake unknown to science. Unlike the naga or ninki-nanka, the sankuni is also associated with landslides. Its underground crawling is supposed to cause massive shifts in wet earth. This sounds much like the weird South American serpentine cryptid known as the minhocão that is said to cause disruption, uproot trees, destroy houses and even alter the course of rivers. The sankuni is not wholly malevolent. Indeed, in legend, it is said to allow humans to use its great coils as a bridge allowing them to cross rivers. It is also said to manifest in dreams warning people of impending landslides. The sankuni is said to crow like a rooster much the same as the crested crown cobra of Africa. Its likeness to both the European basilisk (save in vast size) and the giant serpentine lindorms and worms hardly needs to be stated.

Rudy told me of another weird entity from folklore: the *skaul*. This is a vampiric entity that resembles a normal human being by day but at night its head detaches from its body and flies about as an independent entity. It has luminous hair and saliva. The skaul is said to feed on excrement and rubbish but also to suck up the human life force causing the victim to fall ill, weaken and finally die. The skaul may have been an early attempt to explain disease and illness. The luminous hair and saliva might well be based on early sightings of ball lightning or some other meteorological phenomena. The skaul has analogues across Asia with the Malayan *Penanggalan*, the Philippine *Manananggal*, the Balinese *Leyak*, the Thai *Krasue* and the Japanese *Nukekubi*.

Our meeting with Dipu Marak, the man who had been on the trail of the mandeburung for many years, offset Tura's unappealing nature. A delightful man, Dipu has a deep and infectious passion for the Indian yeti.

He told us how he recalled hearing stories of the beast in his childhood and how they had sparked his lifelong interest. With a Garo mother and a Bengali father, Dipu is a huge fellow who towers over everyone else in the town.

We moved from Tura, in the West Garos, down to Siju in the South Garos where we were met by Rufus - a friend of Rudy's - and another guide. We stopped in a rather down at heel and basic, but clean, tourist lodge. Close by were the Siju Caves where the village Headman had supposedly encountered a mandeburung several years before. The whole area was awash with wildlife from Indian false vampire bats and tokay geckos in the kitchen, to tarantulas on the walls outside. Jon McGowan used some fishing line and a live cricket to go tarantula fishing, baiting the spider out far enough to enable it to be photographed.

The caves themselves were amazing. Apparently, they go for miles with many smaller passages branching off the main cave. Fulvous fruit bats roosted in the cave and bizarre white fungus sprouted up from their droppings. The waters that ran through the cave were alive with tiny fish, shrimp, crab and cave crayfish. A swarm of them were feeding on a dead bat. Jon found two recently demised bats and decided to take them back with him to be stuffed. Huntsman spiders as broad as an adult human hand scurried over the rocks. I was excavating in the earth of the cave in the hope of finding some bone material as Pintu, one of the guides/ porters, found a section of what looked like leg bone under some rocks. It was around six inches long. Upon examining it in the daylight, Jon thought it looked like the femur of a biped. We kept it for analysis.

The paradox of the jungle is that although it contains the greatest concentration of life anywhere on Earth animals are more difficult to see here than anywhere else. Creatures can hear a human coming from a long way off and melt like ghosts into the shadows. Wildlife is much easier to spot in open grassland areas. In all my time in many rainforests around the world I've only seen a handful of large animals. Whilst most of us had been away in the jungle, Dave Archer had stayed by the Simsang searching for snakes and looking for animal tracks. He had found the footprints of a tigress in the sand - it was good to know that there were still tigers in the area.

Later we interviewed the Headman of the village, Gentar. He had encountered something strange in Siju caves several years before, something that had frightened him so much that he refused to go back there. He and some friends had been fishing by the

light of burning torches. They had heard a noise that he described as sounding like someone treading on bamboo. On investigation, they found wet footprints on the rocks, which were human-like but of a vast size. They led down one of the passages that turned off the main one. The group thought a mandeburung had entered the cave from one of its many jungle entrances, and they panicked and fled the cave.

I found it odd that such a creature would be lurking so close to human habitation but I was to hear subsequent stories of them approaching other villages. Cave systems retain a stable temperature, it could be that the creature had entered the caves to keep cool or possibly to hunt for crabs.

We travelled to the village of Imangri and trekked into the jungle beyond, where we saw simulacra of a footprint in limestone beside a river. It is a natural formation, but the fact it has been linked with the mandeburung argues that in order for such an association to have arisen, the creatures must have been known of for a long time. Swarms of yellow butterflies flittered around and we rested awhile beside the waters. Chris, had again felt ill that day and therefore stayed behind in the village.

We returned to Imangri and interviewed the Headman, Shireng R Marak - a 56-year-old with two thumbs on his right hand. In 1978, he and some friends were hunting in the forest, and as it was beginning to grow dark, he heard something big and powerful crashing through the forest. He heard a loud, deep call, which he imitated for us: AUHH!-AUHH-AUHH!. He had heard village elders talking about the mandeburung and demonstrating the sound it made. He and his friends ran into a cave and lit a fire at the entrance., and they heard the creature bellowing and crashing around outside the cave all night. At first light it moved away into the forest and they ran back to the village.

Shireng said that 40 years ago sightings of the creature were more common. His friend's grandfather had shot one. He said it was man-like, covered in black fur and had a face like a monkey.

Tura itself is devoid of anything approaching nightlife. The one bar in the town was at the *Sandre Hotel* and closed at 10 pm sharp. The bartender seemed totally disinterested in making money and resented anyone who entered the bar after 9.45 pm.

The two hotels in Tura both had restaurants that were spectacularly badly run. Their menus were surprisingly varied, but most items on them were not available. This made ordering food a bit like the cheese shop sketch in Monty Python. Far worse than this, though, was the service. On one occasion, we ordered some soft drinks, and an hour later they still had not arrived despite three waiters standing around next to the fridge in which the drinks were. In the end, Dipu himself had to go and open the fridge and point the drinks out to them. On another occasion, I ordered soup and bread - the soup took an hour to come and the bread turned up an hour after that.

We were to spend the next day interviewing a number of people around Tura. The first on our list was Dr Milton Sasama, the Pro-Vice-chancellor of the Garo Hills University, who has written a number of books on the history and folklore of the Garo Hills. However, he does not believe in the mandeburung as he has never come across descriptions of the beast in any of his studies. He has only heard of the monster, like a giant orang-utan, in the past 20 years. He also asserted that there was no tradition of a yeti-like creature in Assam, the Indian state that lies between the Garo Hills and the Himalayas.

Our next interviewee was Llewellyn Marak - the uncle of Rufus - who is a noted naturalist and author of a number of books on the wildlife of the Garo Hills. In 1999, he came across a set of four huge, man-like footprints at Nokrek Peak around 21 km from Tura. They were found beside a stream in sand and were 18 inches long, and led away into the jungle.

Llewellyn's grandfather was a renowned hunter who amassed a large collection of trophies. He had encountered the mandeburung on a hunting trip many years before, when he came across the beast in a jungle clearing. It resembled a huge gorilla and was black in colour, and moved around on all fours giving the impression that it was searching for food. Occasionally it would stop and sit, appearing to eat something. Llewellyn's grandfather became afraid and backed away. This is the only report we have of the creature moving on all fours, but then again it may have been doing this in order to forage for food. The experienced hunter was sure what he had seen was not a bear.

Llewellyn had also heard stories of giant catfish and giant freshwater stingrays, much like those said to lurk in the Mekong River of Indo-China.

Following this interview, we moved on to speak to Rufus' uncle - a surgeon called Dr Lao. Dr Lao also believed that the mandeburung existed but he thought that it was

that no such creatures are reported from Assam.

He writes…

"Singhason peak and some nearby areas are sacred to the Karbis. Here in the dense forest lives the Khenglong-po, the legendary 'hairy wild-man'. The Khenglong-po is an important figure in the Karbi folk tale. Whenever I used to get reports of its existence, I dismissed them as fable or mistaken identification of an ordinary animal. But when the much experienced Sarsing Rongphar gave me a fresh report, I had to re-think. Sarsing had been my guide in parts of the Dhansiri Reserved Forest, and I found him to be an accurate and reliable observer."

Sarsing was a hunter who used dogs to sniff out game such as muntjac and porcupine, which he then dispatched with a long hunting knife. Even before his arrival a Karbi, Along Awaruddin Choudry, had heard of sightings of a large, bipedal ape. At first he asked witnesses if they might be mistaking the creature with a stump-tailed macaque (*Macaca arctoides*) or a hoolock gibbon (*Hoolock hoolock*) but the witnesses rejected this as they were familiar with both species. However, when his trusted guide told him of an encounter with the beast, Choudry was forced to change his mind.

now very rare. Dr Lao had a collection of books on Indian wildlife, and among them was a book entitled, '*A Naturalist in Karbi Anglong*' by Awaruddin Choudry, first published in 1993. The book, by one of India's best-known naturalists, records his time in the Karbi Anglong district of Assam, the Indian state to the north of Meghalaya.

One chapter of Chourdy's book is dedicated to the *khenglong-po*, a yeti-like creature seen in the area. As Assam borders onto Bhutan there is a link, or corridor if you will, directly from the Himalayas down to the Garo Hills along which yetis are reported, which totally refutes Dr Milton Sasama's assertion

It was on 13th May 1992 that Sarsing Rongphar and his friend Buraso Terang took Sarsing's hunting dogs into the Dhansiri Reserve Forest. In the afternoon, they came upon large man-like footprints that were around 18 inches long and 6-7 inches wide.

The pair followed the tracks for 3 kilometres until their usually brave dogs began to panic. Fearing an elephant or tiger was close by they crept cautiously forward, and soon a loud breathing sound became audible - a '*khhr-khhhr*' sound. From 80-90 metres away they saw an ape-like creature leaning against a tree, apparently asleep. The witnesses were at a higher elevation than the creature and had a clear view due to the fact there was no dense undergrowth obscuring their vision.

The creature was jet black like a male hoolock gibbon. It had thick bear-like hair on the body. The hair on the head was long and curly. The creature was a female with visible breasts. Its mouth was open and large, human-like teeth were apparent. The face, hands and feet were black and ape-like. In front of the creature was a broken tree and the hunters thought the creature had been feeding on it. They observed the sleeping animal for around one hour. Sarsing likened it to a giant hoolock gibbon but with much shorter fore-arms.

On reaching their village, they told tribal elders of what they had seen and were informed that it was a Khenglong-po, a kind of hairy Wildman that was thought to be dangerous.

Choudry took Sarsing to his camp and showed him pictures of the Asian black bear *(Ursus thibetanus)* standing on its hind legs, and the mountain gorilla (*Gorilla beringei beringei*). The hunter identified the latter creature as being a Khenglong-po whilst recognising the former for exactly what it was. Choudry interviewed Buraso Terang separately and got the same answers.

A Khenglong-po was once supposed to have wandered up the railway track from Langcholiet to Nailalung.

On another occasion Choudry talked to some hunters from Karbi Anglong in central Assam. They spoke of a large, herbivorous, ground-dwelling ape that they called *Gammi*. According to them two Gammis were seen together in 1982 feeding on reeds on the eastern slope of the Karbi Plateau in the upper Deopani area. An elderly hunter had encountered one in the Intanki Reserve Forest in Nagaland in 1977-78. The creatures are said to be covered in grey hair and to be man-like in appearance. The name Gammi means 'wild-man'.

Choudry concludes...

> "It seems possible to me that a terrestrial ape, larger than the gibbons existed in some remote parts of Karbi Anglong and adjacent areas of Nagaland. The creature was always rare and preferred the remotest corner of the jungle, and, hence, evaded discovery by the scientific world. Now with the forests vanishing everywhere, this ape perhaps faces extinction. Expeditions to the heart of the Dhansiri Reserve Forest and Singhason area may well produce some result. But for now, I am looking for any fossil evidence including skull, bone or part thereof. This will at least put the Khenglong-po at its right place, even if it is extinct. Lastly, if a large mammal like the Javan or smaller one horned rhinoceros (*Rhinoceros sondaicus*) can be discovered in recent years in a small pocket of the war-ravaged Vietnam, outside its known locality in Indonesia and beyond anybody's expectation, one cannot rule out a Khenglong-po in the forests of Karbi Anglong."

We can see, then, an unbroken link of yeti sightings from Bhutan down into India.

The following day we interviewed another witness. He was a 51-year-old teacher called Kingston. In 1987, he and a friend were on Tura Peak, when he saw large, five-toed, man-like tracks in wet sand beside a stream. The toes and heels extended far beyond his own and sunk an inch into the sand whereas Kingston's own tracks only sunk in half an inch. My size nines were bigger than Kingston's, but he told me that the creature's tracks were bigger than my feet. He also heard the mandeburung's cry, AUHH!-AUHH-AUHH! He imitated the sound, which was in line with that made by other witnesses. He had wanted to investigate further, but his friend was too afraid. Kingston added that he has heard the cry on Tura Peak since, within the last few years.

The next day we met with a most impressive witness in the village of Ronbakgre. Teng Sangma had heard that in April of 2004 a village carpenter had seen a female mandeburung suckling an infant in a bamboo forest close to Rongarre. He did not believe the story but then on the 24th of that month, he and a friend were hunting for jungle fowl in the forest when they came across a huge figure sitting with its back to them. Even in its sitting position it was five feet tall. It was covered with dark hair and had longer hair on its head that fell down onto the shoulders and the back. The shoulders were very broad. It was a female, and was suckling a youngster whose legs were visible at the side of its mother suggesting that the infant was sitting on her lap. The youngster was making gurgling noises. The adult was pulling down large bamboo stems and plucking off the leaves to eat them. The men got to within 50 feet of the creatures, watching them for 2 minutes before becoming afraid, when they backed away leaving the creatures, which – apparently - had not noticed them.

The following day we met another impressive witness. Nelbison Sangma was a farmer from the village of Sansasico, and he had observed a mandeburung for three days running in 2003. Nelbison was some 500 metres from the creature, and was able to look down upon it as it was on top of a smaller hill. When he first saw it, the mandeburung was standing under a tree and Nelbison told us that it was nine feet tall and covered with black hair. Whilst he watched, it moved around for an hour and then slept in a nest it had constructed by pulling down branches much like a gorilla does. The next day the creature was in the same place and appeared to be sunning itself, and this time he watched it for half an hour. On the third day he saw it again and it was wandering about and foraging.

The following day he took some other villagers to the area and showed them the nest and they noticed there was a monkey-like smell that pervaded the surroundings. They

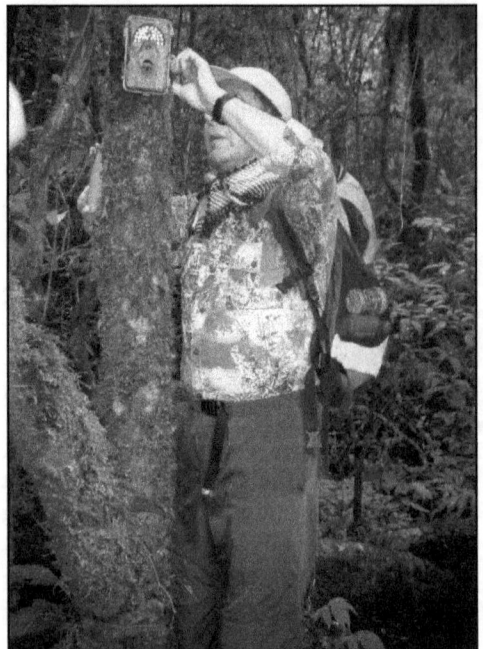

found man-like tracks 18 inches long and a huge dropping which contained banana leaf fibres and was the length of a human forearm.

We returned to Tura, and spent the next day chasing up some leads and loose ends. We photocopied the relevant chapter from Dr Lao's book and then visited the rather shabby library to see if there was anything on the mandeburung and the sankuni in any of the books there. We turned up absolutely nothing there, so we tried to track down Albin Stone, the man who was said to have eaten the flesh of a dead, juvenile sankuni, but he was not at home.

Dipu showed us a rib bone found by his father at Balpakram in 1989. I thought it looked more bovid than primate, but we took a sample from it. Dipu also showed us a collection of hairs he found at Nokrek in 2006, which looked to me like goral (*Naemorhedus goral*) a goat-like antelope known to inhabit the area. All of these, together with the muntjac sample and the bone from Siju Caves, will be sent to Lars Thomas and his team for analysis.

We met Dipu's uncle - Garfield - who, whilst fishing in 1956 or'57, came across a mandeburung print beside a stream. It was on a rock and formed by water where the creature had recently walked out of the stream and across the rocks.

Garfield also claimed to have seen the trail of a sankuni in the 1970s. The track emerged from the Garo River and went for 100 metres under a wooden bridge, destroying some of the supports, and then crossed a paddy field and entered a marsh. However, upon closer questioning Garfield told us that there were *two* tracks running parallel to each other and

the ground/vegetation between them was undisturbed. This sounds very much like the tracks of some kind of all-terrain vehicle rather than a giant snake that would leave one single furrow.

All too soon, our time in the Garos was over. We had to say our goodbyes to Dipu, Rudy, Rufus and the others, before we returned to Delhi where we had a day to see the local sights. We had intended to take in the Taj Mahal but it was too far away, so we made do with the local sights such as Humayun's Tomb and the Qutb Minar. Ironically, we saw more wildlife around the city than we did in the jungle. There were flocks of ring-necked parakeets (*Psittacula krameri*), troops of rhesus macaques (*Macaca mulatta*) and Indian three-striped palm squirrels (*Funambulus palmarum*).

I am convinced that the mandeburung exists and that it is one in the same as the larger kind of yeti. The best model we have for this animal is a surviving form of *Gigantopethicus blacki*. As for the sankuni, its startling resemblance to the Indo-Chinese naga, the West African ninkinanka, the Central African crested crowing cobra and many other monster serpents, convinces me that there are more to these stories than hot air. Already there is talk of returning to the Garos in a few years time, probably to mount an expedition down into the gorge at Balpakram. Kipling's India is still alive if you look hard enough and I intend to return there.

This article has been distilled from the India Expedition report, available for £9.99 from CFZ Press. There is a full account of the expedition and over 100 photographs.

STRANGE
N E W S
FROM
IRELAND:
O R,
A true and perfect Relation

Of a famous

F I S H

Taken at *K I N G S A L E.*

ON the 26th of *July* last, one
Thomas Davis and *Andrew*
Simpson being at Four a Clock in
the Morning about their ho-
nest

A 2

nest vocation of Fishing , in
which they laboured an hour to
no purpose ; but at length they
perceived afar off somewhat
move in the water, of an unusual shape and bigness ; upon
which they made unto it, but as
soon as it saw them it dived
down into the water : so that
after a tedious and vain search,
they returned back to the
Town and informed one Mr.
Rocke, a responsible Inhabitant
there , of what they had seen ;
who with Three of his Servants
came forth with Muskets char-
ged, one standing on the shore,
and the other two in the Boat
with the Fisher-men , with
whom also was Mr. *Rocke*. They
had

had not long rowed about the River, but they again perceived its head popping up and down, at which they difcharged their Pieces, but to little effect : For it dived again out of their view, and did not appear in an hours time after; but upon its rifing again they fhot it into the back; whereupon it made with all pofsible fpeed towards land, where (though with great difficulty) it was taken.

The defcription of its parts is as follows.

On

On the head of this wonderful Creature (which can be no
nearer refembled to any thing,
than the head of a man) was
long black Hair : its Face had
the exact fhape of a Lion ; fo
terrible and grim, that it ftruck
terrour on all that beheld it.
It had two fore-feet like thofe
of a Bull, cloven ; its hinder
feet being like unto an Eagles
Talons , with very long and
fharp Nails ; fo that where Nature commonly orders Fins,
there were perfect Feet of o-
ther Creatures of a clean con-
trary Element. His mouth was
guarded with three long fharp
Horns , with which when he
was

was on Land he fo dangeroufly
wounded one of the forefaid
Mr. *Rock's* fervants in the thigh,
that he remains very ill of the
fame, infomuch that his Reco-
very is much doubted by all,
notwithftanding the Advice
and Afsiftance of the moft skil-
ful Chirurgeons in thofe parts.
On its Back was the perfect re-
femblance of an Hour-glafs,
and the appearance (as fome
fancied) of a Spade and Deaths
head, to the amazement of all
its Spectators.

It is in length Twelve Foot,
and Five in bredth.

On each fide its Breft are two
Paps, like unto thofe of women.

Its

Its Belly is ſmooth, being be-
ſpeckled with ſpots of divers
colours.

When this Letter came away
it was not quite dead, though
it had been much wounded;
but is with much ado brought
into the aforeſaid Town of *Kin-
ſale*, where it now remains the
Wonderment of all its numerous beholders.

F I N I S.

This pamphlet on a Strange Fish off Ireland in 1677, comes via Richard Muirhead who asks
that we thank the staff of Lincoln Cathedral Library for providing the pamphlet in the summer
of 2009 and thanks to Nic Moran for his help.

KALLANA – ON THE TRACK OF ALLEGED PYGMY ELEPHANTS OF KERALA, INDIA

Matt Salusbury

Round about 2005 I became aware of news reports in the South Indian regional English language press on "kallana" – a mystery pygmy elephant sighted in and around the Neyyar-Peppara Wildlife Sanctuary in Kerala, India. These elephants are supposed to have a height in adulthood of 5ft (1.5m) at the shoulder – or less. Some reports put them at 4 ½ ft in adulthood, and say they are particularly nimble, scrambling over rocks at great speed. (Conventionally-sized adult Asian elephants on the Indian sub-continent start at around 7ft at the shoulder and up.)

There would be an occasional report in local editions of *The Hindu* or *Deccan Herald*, being picked up Indian news agen-

Dead female "kallana" found by Sali Palode and Mallan Kani near the shore of a lake in Neyyar-Peppara Wildlife Sanctuary, March 2005. Certified as a dead, young, elephant by the local Wildlife Warden and then cremated *in situ* following Kerala Forest & Wildlife Department best practice.

cies. A March 2005 sighting involved a group of four kallana, and there appeared in *The Hindu* a photo of an alleged kallana standing in a forest clearing, with – frustratingly – nothing to give any indication of the scale. At the same time, a dead female kallana was found by a lake in the Sanctuary, although no photos of this emerged.

In April 2010 more photos appeared – of a lone "tusker" (male with tusks, conventional Asian elephants don't always have them) standing on the shore of the same lake. There was one common factor in all the sightings since 2005 – art teacher and amateur photographer Sali Palode was always one of the people photographing it, and Mallan Kani, of the Kani "tribal" group that lived in the forest was always his guide.

I researched pygmy elephants for a *Fortean Times* article which can be found at (http://mattsalusbury.blogspot.com/2009/10/pygmy-pachyderms-on-track-of-pygmy.html) My researches took me to the British Library, to early 20th Century French language sources that Heuvelmans had cited on his *On The Track of Mystery Animals* chapter on "water elephants" of the Congo.

I found these sources were rehashes of slightly earlier reports. An examination of these showed the original reports had be-

Male "tusker" spotted on the shore of a lake in Neyyar-Peppara Wildlife Sanctuary, April 2010. Sali, Mallan and two other amateur wildlife photographers were present. Very experienced photographer Balan Madhavan estimated from metadata on focal length that was attached to the photo file that the "object" (the elephant) was around 100ft away from the camera.

come somewhat embellished by the time they came to Heuvelmans' attention. Heuvelmans hadn't mentioned pygmy elephants in Asia at all.

I also went "backstage" at the Natural History Museum, London, where researcher Victoria Herridge showed me the Bate Collection of fossil pygmy elephants of the Mediterranean islands. They were much older and very, very different than what was being claimed in the Congo and in Kerala – the "water elephants" and kallana were supposed to be just smaller but otherwise very similar versions of conventionally-sized elephants. The pygmies of the fossil record were something else – comparatively bigger heads, shorter limbs, simpler teeth, possibly a shorter life span and giving birth to "litters" rather than having just one calf at a time. Surprisingly, Herridge didn't completely rule out the possibility of a five-foot pygmy elephant in either the Congo – where she'd studied living forest elephants – or Kerala.

Kallana bothered me. The evidence for it seemed better than the frankly dubious pygmy elephants of the Congo. Professor Sukumar Raman, *the* expert in Asian elephant ecology for the whole world, also seemed to have an open mind on kallana, and his Indian Institute of Science took it seriously enough to put together a team to go look for it in 2009, only cancelled through heavy and unexpectedly early rains.

I was also intrigued by the fact that the Neyyar-Peppara Sanctuary wasn't exactly at the ends of the earth, or only reachable after months of trekking through wild and inhospitable jungle. Kallana's alleged habitat was

– as I was to experience myself – only an hour's ride by motorcycle rickshaw from Kerala's state capital and international airport at Trivandrum. It was the equivalent of having a mystery animal in the South Downs just outside Brighton. Compared to the cryptids supposedly living in the wilds of Mongolia or up the Himalayas, kallana was right under our noses.

There's a time when you've spent enough hours in the library on armchair cryptozoology, and when the only way to take such investigations further is to get off your arse and go take a look yourself. So in March this year I took a couple of weeks off work and went out to investigate these kallana reports.

In Kerala's capital, Trivandrum, I encountered several carvings and statues of what were described as "unicorns" or "elephant dragons" – horse-bodied, eagle-clawed beasts with elephants' heads. Some grasped their trunks in their talons, some had trunks reaching down towards considerably smaller "baby elephants" whose trunks reached up to theirs. Some had small crests or tufts on their heads. Some had multiple tusks growing out of the sides of their mouth where their teeth should be, like the mouthparts of a monster prawn. They had a Chinese or even Indonesian look to them, and the elephant bits were anatomically very accurate.

These "elephant dragons" were in the Maharaja of Travancore's 18th Century palace, in the huge temple next door, and guarding other temples nearby. I talked to the palace and temple guides, who told me the "elephant dragons" were carved by wood

OPPOSITE ABOVE: Sali Palode (left) and his guide Mallan Kani (right) trekking in Neyyar-Peppara Wildlife Sanctuary.
OPPOSITE BELOW: Mallan (left) and Sali (right, with backpack) climb the steep elephant paths in the thick forest

"Makara" - "elephant dragon" guarding the main temple in Trivandrum, Kerala, added early 18th century. Makara's trunk reaches down to join the trunk of a "baby elephant".

OPPOSITE: "Makara" - "elephant dragon" at the Maharaja of Travancore's palace, early 18th century. Carved by artisans brougt in from Tamil Nadu.

and stone carvers from Tamil Nadu in the late eighteenth century, during Travancore's zenith. The stone dragons were added to the much, much older temple at that time. In Bangalore, I saw that the coat of arms of the surrounding state of Karnataka also has "elephant dragons."

Sukumar told me these are "makara", and they're not exclusive to India. They seem to be an architectural flourish or heraldic beast, at a time when Travancore (South Kerala and bits of Tamil Nadu) were establishing diplomatic relations with European powers.

Within the CFZ, Richard Freeman told me "makara" resembles "bakau", a stylised tapir of Japanese legend. And Oll Lewis told me that many of the magical creatures of Chinese traditions started out as conventional animals but were deliberately embellished and augmented with fantastical bits to create unique heraldic animals to identify noble families. Something like this may be at work with "makara."

I don't know how the existence of "makara" – a mythical elephant dragon often associated with a small elephant – fits in with the tradition of kallana. Sukumar says there's been little by way of research into the origins of "makara". I invite any art historians who don't mind getting up really early (Trivandrum is 5 ½ hours ahead of GMT) to make calls to India to investigate further.

No, I didn't find any pygmy elephants, nor did I see any conventionally-sized wild elephants. I did interview art teacher and multiple award-winning amateur photographer Sali Palode and his "tribal" guide Mallan Kani, of the forest-dwelling Kani people. They have been tracking "kallana" for over a decade, and have had three sightings in that time. They were able to photograph kallana on two of these occasions, in 2005 and 2010. (Sali speaks Malayalam only, and his agent Balan Madhavan interpreted for me. I hope to have an extract of the interview linked from my blog shortly www.mattsalusbury.blogspot.com.)

I also had a chance to see some young captive elephants having a bath at Kodanad Elephant Camp, a Kerala Forest and Wildlife Service's facility for working rescued elephants and their mahouts. It was good to get up close to them, and to compare young conventionally-sized elephants with the photos of "kallana."

Sali seemed completely sincere. A hoaxer he was certainly not. I did note, however, that his description of the distinct characteristics of kallana omitted some of the features referred to on his website – a longer body and a bigger head, for example. Sali speaks very little English, and he's dependent on a friend who's a naturalist for translations on his website. It may be there's been some embellishment by the time his description reached his website.

Some features that Sali said were proof that kallana were adults were features I'd noticed myself on captured elephants taking a bath in the river Periyar that I saw while staying at Kodanad. Six or seven-foot juvenile female elephants – they couldn't have been more than seven or eight years old – had the nipples, wrinkles and the long tail that Sali thought were adult features.

I'd also read a lot about how the elephants of different parts of India are different. In Kerala, they say "their" elephants have a massive head, while in Bihar, elephants are supposed to be hairier, and so on. Both Prof Sukumar and Prof Joseph Cheeran, a retired elephant vet, told me that what people in different parts of India "know" about their elephants (a source of regional pride) has a status roughly equivalent to something you heard from the man in the pub. The only regional variation Sukumar could discern was that the radio collars he had made for Kerali elephants wouldn't fit on the Bihari elephants as they had slightly thicker necks.

We imagine that elephants make a lot of noise and make the ground shake when they walk. But when I was watching young elephants being bathed in the Periyar river, the only indication that a much bigger but still young male elephant – easily an eight-footer at the shoulder – was right behind me was when one of the mahouts gently tapped me on the shoulder and asked me to get out of the way. The elephant moved completely silently.

Sali and Mallan took me into the Neyyar-Peppara Sanctuary to show me the places where they encountered kallana. It's thick forest with steep, single-file paths up and down the hills. The paths are elephant tracks, as evidenced by the dung piles with mushrooms growing out of them. The places where Sali and Mallan made their sightings were all on the shores of a small lake at the edge of the forest.

We did have a close encounter with a herd of about 20 gaur (wild forest bison) that Mallan found for us. Mallan's forestry skills are impressive – he suddenly said, "Gaur! Guar!" although Sali and I saw and heard nothing, and he then disappeared into the forest. Ten minutes later a herd of gaur came stampeding straight at us. Some say that "kallana" are just young elephants playing a short distance from a herd that's unseen and close by, but if the herd *were* close by, Mallan would know about it. I had the rare privilege of coconuts for lunch in one of the Kani hamlets in the forest after our trip. Access to "tribal" areas is normally restricted. We had cleared it with Sharma, the Trivandrum Division Chief Wildlife Warden, who knows Sali well.

After my one trip into Neyyar-Peppara I was back in the capital, Trivandrum (it means "Holy Snake City") after two very short bus rides, in time for lunch. I could have gone into the Sanctuary again, although it would have been very difficult to arrange – Chief Wildlife Warden Sharma had only given me permission to go in condition I was accompanied by Sali and Balan. Sali had taken a day off work from his art teacher job at the government school at short notice, and wasn't able to arrange any more leave. This being India, I'd spent several days of my limited trip waiting around in Trivandrum while Balan tried to get Sali on the phone, it looked at one point that it had been a wasted journey and I wasn't even going to meet Sali. And even if I'd gone back into the Sanctuary, let us not forget that Sali and Balan have been tracking kallana for at least a decade, and had only three encounters and one dodgy sighting of its dung (see below.) The chances of me tripping over kallana – or even a conventional elephant – in the thick, thick forest, were never great.

I have purchased from Sali's agent licenses to use a couple of Sali's photographs, including one never published of a dead female he found by the lake in 2005. The local wildlife warden certified it as dead, stating it was a young elephant, and it was quickly cremated in line with Forest Department practice. Regional newspaper reports at the time saying a DNA sample had first been taken were incorrect. As per the licence, I have included some very small, low-resolution versions of the photos with these articles; to stop you lot nicking them. Sali has to make a living.

As Prof Cheeran (see below) noted, "every newspaperman wants their scoop." Some kallana reports may have been exaggerated by local journalists wanting the make the story more interesting, and who were writing up witness testimony in a different language (English). It's noticeable that the sudden interest in kallana around 2005 coincided with the aggressive promotion of the "pygmy elephants" of Sabah – the northern tip of Borneo, in Malaysia territory – by the wildlife tourism industry. The Borneo pygmies are on average only six inches shorter than their mainland counterparts. And there are Keralis working in Malaysia who presumably told the folks back home. (One in five adult male Keralis work abroad, mostly in the Gulf.)

I interviewed two elephant experts. Prof Sukumar Raman is *the* expert on Asian elephants for the whole world, and I flew to Bangalore to talk to him on the huge Indian Institute of Science campus. The IIS grounds are so huge they have their own airstrip. On the taxi ride out of the campus, I saw what appeared to be a steampunk Nehru-era particle accelerator being dismantled, which resembled something out of the *League of Extraordinary Gentlemen*. I also met and interviewed Prof Joseph Cheeran, a vet and an expert on captive elephants, in the Kerali city of Thrissur. We'd made contact by email and brief phone calls, but when we met face-to-face we found we had real difficulty understanding each others' accents!

I have to say my initial conclusion is I am more sceptical about kallana than when I first arrived in Kerala. Prof Sukumar and Prof Cheeran said that many of the unique characteristics that Sali says distinguish kallana from conventionally-sized Asian elephants are perfectly consistent with young Asian elephants, and there is a big range of size, tusk development and behaviour in young Asian elephants. All the kallana sightings were in the dry season – Kerala misses out on one of the monsoons and has a longer dry spell – giving the elephants an emaciated appearance, which could be what Sali and Mallan were seeing.

It turns out that kallana doesn't go back all that far in time either. Sali said Mallan first drew his attention to kallana "twenty-five years ago" on the summit of Agasthya Mala, south Kerala's tallest peak, when he saw piles of smaller-sized dung. Prof Sukumar pointed out that younger elephants produce smaller balls of dung. Sukumar is also from South India, and said that "twenty-five years ago" in South India doesn't necessarily mean 1986, but "a very long time ago," so long that you can't remember.

Sukumar also told me that the stories of kallana first arrived in Kerala around the same time, with reports coming over the border with the neighbouring state of Tamil Nadu to the East. (Neyyar-Peppara goes up to the Tamil Nadu border.) I wasn't able to communicate directly beyond single English nouns with Sali and Mallan on our trip into Neyyar-Peppara, as we didn't really have a language in common, but it's no longer clear to me whether kallana is an ancient tradition of the Kani, or a secondhand report acquired via local media by the Kani and others in recent times from Tamil Nadu. There are some Kani settlements (and a lot of other forest-dwelling "tribal" settlements) in Tamil Nadu, but the Kani in Kerala seem to be isolated from and unaware of the Kani in Tamil Nadu, who speak a different language. More study of the transmission of kallana reports is needed.

Sukumar was more open to the idea of kallana, and said that, as a scientist, "I wouldn't rule it out." He suggested that kallana is an example of "phenotypic plasticity", varia-tions within any given population, and felt the most likely explanation is that there's a family group of slightly smaller than usual individuals in the sanctuary. Sukumar pointed out that being smaller would be a good adaptation in negotiating the thick forest slopes of Neyyar-Peppara. Sukumar has studies the elephants of Burma, that also live in thick forests, and they are smaller than Indian elephants, but definitely the same species.

While Balan Madhavan, Sali's agent, was emphasising to Sali the need to gather dung, hair and other discarded bits of kallana for DNA analysis, Sukumar said not to bother, as any variations unique to the alleged "kallana" family of smaller elephants probably wouldn't show up in the DNA, as they're well within the range of what you'd expect in conventional-sized elephants. Sukumar said my visit had reminded him that he really ought to do something about getting the IIS on the next Kerala Forest Department elephant survey of Neyyar-Peppara with a view to looking for kallana – his considerable academic commitments permitting. His team was rained off the last one in 2009, the next one is in January 2012.

Sali also claims to have seen a mystery tree crab in the Neyyar-Peppara reserve, living in gaps in trees that are not particularly anywhere near water. There's a photo of a tree crab on his website, erroneously filed on the "insects" page. (I've yet to get in contact with the naturalist who runs Sali's website, www.salipalode.com.) Sali drew

ABOVE: We didn't see any elephants, but we did have a close encounter with a herd of gaur (wild forest bison) in Neyyar-Peppara Wildlife Sanctuary. Sali photographing at right.

BELOW: Forest tracker Ballan Madhavan recreates his 'kallana' sighting, standing at the same spot it was observed on the shore of a lake in Neyyar Peppara Wildlife Sanctuary.

me a sketch of the tree crab, which looked very different to the photo, more like a spider. He also said he'd seen exotic tarantulas in the sanctuary, although it's not clear whether he was saying they were unknown species. I hope to delve more deeply into the Neyyar-Peppara tree crab mystery. On another backstage trip to the Natural History Museum, I was shown a photo of one of those Australian spiders that weaves nets that it throws at its prey, and it looked awfully like the Neyyar-Peppara tree crab drawing.

Balan Madhavan, who's a well-known wildlife photographer, said he hoped to get a photo of another Kerali cryptid, *pogeyan*, the grey clouded leopard. He's spoken to foresters who've seen it, and is convinced it's for real. Pogeyan's alleged range is in the far north of Kerala, in the tea plantations of Malabar.

"Makara" mythical animal tie-ins aside, while kallana doesn't seem to be a genuine pygmy elephant, it is evidence for something important at work. As already mentioned, Sukumar pointed that the Asian elephants of Burma - while most definitely very closely related to the elephants of India and practically indistinguishable through their DNA – are slightly smaller, because being slightly smaller is an adaptation for survival in thick forests.

British-born ivory poacher, big game hunter and all-round cad turned "conservationist" W Robert Foran (*The Field*, November 18 1950) concluded that the pygmy elephants of the Congo were a

scam, but accepted that in West Africa there was an "undersized 'race'" of elephants in really thick forests. (Today's scientists would call this an "ecotype", while Heuvelman's Francophone sources called it a "*race ecologique*". Laypersons would call it a "breed" of elephant.) Climbing with Sali and Balan in single file up and down some narrow paths on breathtakingly steep slopes in Neyyar-Peppara, I came across some fresh elephant dung, and it was only then I realised these were elephant paths. Being ever so slightly smaller in Neyyar-Peppara would surely be a useful adaptation in this cramped environment too.

Kallana would seem to be part of a "breed" or even a single family group of elephants that have got ever so slightly smaller to adapt to a very thick forest. While I saw elephants trenches dug in the forest to keep the elephants out of the Kani villages, the forest has only been isolated for well under 30 years – nothing like enough time for a new sub-species to emerge.

Sukumar pointed out that, among all the now isolated elephant populations of India, what's surprising is how much healthy genetic variation there still is among these populations. He should know, he was personally involved in most of the DNA surveys of Indian and Burmese elephant populations. Any unique characteristics among the slightly smaller Neyyar-Peppara elephant individuals is unlikely to show up in their DNA, and will most likely breed out through genetic contact with the other elephants that go in and out of the park. (There were 21 conventionally-sized elephants identified in the last Kerala Forest and Wildlife Department survey of Neyyar-Peppara in 2008,

although Sukumar pointed out that there could well have been some more that had temporarily gone over the other end of the mountain in Tamil Nadu, or up North into the Western Ghats, possibly almost as far as Goa. Most such survey results are based on counting piles of dung; you don't get to *see* many elephants in such a thick forest.)

Regarding kallana's nimble feats of scrambling rapidly over rocks, which Sali said he'd seen, John Hutchinson of the Royal Veterinary College, who's studied elephant locomotion and whether elephants can run, told me, "juvenile elephants are relatively more athletic than adults." He suspected any elephant seen nimbly trotting over rocks in this way was probably a less stiff-jointed juvenile.

One other avenue where I totally failed to make any progress while in Kerala was in tracking down naturalist P. Easa, who several people mentioned as investigating kallana around 2000 and giving up after finding no evidence for its existence. He retired and moved away from Thrissur, and I had no luck in following up various leads, including subsequent work as a trustee of a national Indian wildlife charity. My very early morning follow-up phone calls yielded no results – this is, after all, India. Day job commitments prevent me tracking down P. Easa for the moment, but any enthusiasts who are also early risers and want to give it a go, please get in touch for some leads.

There will be more detail in my forthcoming CFZ book, *Pygmy Elephants*, when I finally get around to writing it.

Opposite: Sali Palode (left, with camera) and Ballan Madhavan (right) stand at the spot where they encountered a dead female "kallana" in 2005

Letters to the Editor

The Editor and his band of merry men (and women) welcome an exchange of correspondence on any subject of interest to readers of this magazine.

We reserve the right to edit letters and would like to stress that opinions voiced are those of the individual correspondent rather than being necessarily those of the editorial team or the Centre for Fortean Zoology. Every attempt is made not to infringe anyone's moral rights or copyright, and we apologise if we have unwittingly done so.

The Snow-Hen of Austerlitz

Hi Jon,

I thought it was about time I shared an encounter with a cryptid with you. I have often told people about what I saw but get ridiculed for it. Anyway when I was a teen of about 14 -16 (I'm 43 now) I was walking my dog up a hill in Lancaster, houses were to the left of me and a field was to my right. It was a dark night around 10 o'clock when 10 feet away from me I noticed a hedgehog crossing the road.

I know it was a hedgehog as I'd seen many in the area, I also had perfect vision and also being an artist I am very observant.

About half way into the road the hedgehog suddenly stopped moved in an erratic way and then did a couple of small bounces, no more than an inch off the ground, then it shot up into the sky where I lost sight due to the darkness and trees hanging over.

Both myself and my dog were alarmed and so baffled we went home.

Now I have told this tale many times and as you can imagine much laughter follows and accusations of what was I drinking or smoking is often asked.

I was totally clean of any intoxicant. I was prompted to write to you as I had a particularly derisive response last night during a party and I thought you would be able to shed some light on whether there have been other encounters with flying hedgehogs.

My response to people who don't believe me is that why would I make up such a story as it paints a negative portrait of myself. I know what I saw and I will continue to tell the story but maybe only to people like yourself who understand and have knowledge of such strange encounters.

Thanks

Richard Disley

EDITOR'S REPLY: Do I know other stories about flying hedgehogs? The answer is an unsatisfactory 'well sort of'. About 15 years ago an April Fool's Day hoax got out of hand when several papers printed accounts of 'flying hedgehogs' (called *Tizzie Wizzies*) and various people who should have known better hypothesised that they were a colony of Asiatic flying squirrels that had escaped from somewhere.

It was soon announced that it was a Beatrix Potter related hoax dreamed up by the local tourist board, and the matter was soon forgotten.

Or so we had thought.

On Friday 19th November 2010 the *Cumberland and Westmorland Herald* published an account of a meeting of the Lakeland Dialect Society which included this passage:

> Mr. Atkinson began by saying he had been born in Langdale and had spent all his life in and around Windermere. His father had been one of the boatmen on the lake, and he recited the nicknames by which several of them were known. He went on to explain that the Tizzy-Wizzy is a fictional creature created by the Windermere boatmen as a tourist gimmick. It was described as being like a hedgehog with wings. The boatmen marketed their evening cruise as a Tizzy-Wizzy hunt.

When Mr. Atkinson left school he worked in a local nursery where he discovered that he had a feel for Latin; and he told the audience how he would go to the Barrow archives and read the Latin manuscripts from Furness Abbey, where he had been excited to find a reference to the Tizzy-Wizzy!

The document, he said, recorded two monks who had witnessed an old blind hedgehog seeking a mate. After having had encounters with a tuft of grass and a brush head, he chanced upon a buzzard that had been grounded. The buzzard nested on Claife Heights and hatched two eggs, which hatched to reveal creatures with spines, black noses and wings.

These creatures fell out of the nest, rolled themselves into a ball and bounced down into the lake. There their diet resulted in a great deal of flatulence, and this could cause them to travel at 70mph on the lake no speed limits then!

Enterprising hoteliers decided to offer rewards for the capture of Tizzy-Wizzies, and these were used in the hydropathic baths to create bubbles creating the first Jacuzzi.

The resulting smells, however, did not prove popular, the creatures were released and are now rarely seen.

A second tale described the first circus to visit the area. Some local farmers saw the strange new creatures and thought that if they crossed a Herdwick sheep with a giraffe it would be able to graze the crags much more easily. This would prove unprofitable since the cheapest cut of mutton is neck end, of which there would be a lot.

The circus then moved on to Langdale where residents were perturbed to find an elephant pulling up all their vegetables. One man reported the occurrence to the police, saying that a strange headless creature was pulling his cabbage up with its tail.

As far as your peculiar experience is concerned, no-one in the CFZ office has heard anything like your story before, and so (like Fortean Esther Rantzens) we must throw the gauntlet down to the A&M readership to see if anyone can come up with a halfway plausible explanation.

Yours,

Jon Downes

1819, The Year Without a Summer

Dear Jon,

I just watched the movie on youtube that you took of the Irish lake monsters. It is a very interesting film indeed.

My name is Luke Fullenkamp from the United States and I am a retired social worker and writer. (You can look me up on amazon.com. I write under the name Luke Steven Fullenkamp). I also have a great interest in cryptozoology. I have studied lake monsters for years--as an amateur of course--and your film is one of the most interesting I've seen in a long time.

I've always held to the theory that these lake monsters were more than likely descendents of the plesiosaurs, having survived the great extinction of the dinosaurs.

After all, many plesiosaurs inhabited inland lakes and seas. I heard you speak of your giant eel theory and you may very well be right; although the film seemed to show that the dark, fast-moving animal was flying through the water with large flippers.

The flippers seem to me to be more like a plesiosaur than an eel, but the view was so distant that I might be seeing something that's really not there.

In any event, I wanted to write and let you know that I enjoyed the film. If you happen to come across any other films or photos that you think are interesting, please feel free to e-mail me the links to them. I'm always interested at looking at new photos or footage of lake monsters.

Thanks again for the look-see. Sincerely,

Luke Fullenkamp

Cage in a Cave

Dear Jon

Watched on the track 44 tonight Jon and

thought I would ask my dear old dad if he had heard of a Great Bustard reintroduction on Salisbury plain before 1970 and he said he seems to remember reading in the 1960's that they tried in the 1950's, but it was a disaster, he can't remember exactly which year, as it's stretching the memory a bit, but he seems to think the birds came from Finland on this attempt as opposed to the Russian tundra.

My father has studied Ornithology all his life and is very knowledgeable on the subject so this may well be right but we are testing his memory a bit.

Hope this is of use.

Carl Marshall
Stratford Butterfly Farm

EDITORIAL REPLY: For those of you who don't watch *On the Track* (and if not why not?) we recently visited *The Hawk Conservancy* in Hampshire, where we filmed their great bustards.

During this segment I mentioned that when I was a boy my maternal grandparents took me to the museum in Salisbury on several occasions during the 1960s, where I saw their stuffed specimens. I was only in England for short periods during during 1963/4, 1966, 1967, and 1969, and I cannot pinpoint the occasions further.

However my grandmother told me that there was an active reintroduction programme taking place, which is strange as the Great Bustard Trust didn't start operations until 1970. Is there anyone else who may have knowledge of any pre-1970 reintroduction programmes?

REVIEWS

Moa Sightings by Bruce Spittle
Hardcover: 448 pages
Publisher: Paua Press Limited; 1st edition (January 1, 2010)
Language: English
Vol1
ISBN-10: 0473153564
ISBN-13: 978-0473153564

Vol2
ISBN-10: 0473153572
ISBN-13: 978-0473153571

Vol3
ISBN-10: 0473153580
ISBN-13: 978-0473153588
http://www.pauapress.com/

Once in a while, I get my hands on a book which makes me think "why on earth don't more people write like this?" A lot of so-called cryptozoological books are, basically, rubbish.

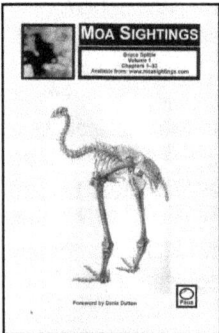

Yes, there are a number of great books out there, but an awful lot just rehash previously covered stories, or dive into the paranormal and sensationalism to increase sales. This book is about as far removed as one can get from these works of tripe.

Split across three volumes, "*Moa Sightings*" is a real behemoth which, due to its high price tag, will sell very few copies. But anyone with a great interest in moas or New Zealand's cryptozoology should sell a portion of the family silver and buy the whole set. In five words, the three volumes are incredible.

Hardback with full colour printing throughout, the books are of an extremely high quality indeed. The front cover, I feel, looks a little basic, but that really does not matter. Upon opening the book, you are greeted to a wonderful range of maps (there are probably hundreds of maps between the three volumes), portraits and biographies of most of the witnesses (those that can be traced anyway), drawings of skeletons and huge numbers of photographs of the areas in which the sightings occurred. This is a book which really draws you into New Zealand. The maps and photographs help you picture the areas vividly in your mind; whilst the long and detailed discussions of each sighting help you assess the circumstances in which the sighting event occurred. The biographies allow you, dear reader, to form an idea about the quality of the witness with a large quantity of unbiased information.

The analysis of the Freaney photograph for instance is 283 pages long. That is enough for a book alone on the photograph. Though the analysis is generally very thorough, and the photographs clear, I would have liked to have seen more photo measurement analysis, perhaps comparing the measurements and angles to a deer. But really, it does take a 'scientific' mind to do analysis like this. Some people enjoy doing such things (like myself), others prefer collecting sighting reports, and others enjoy analysing the wider aspects of the reports. Cryptozoology relies on people from all three camps being present, and Mr. Spittle's books certainly fall into the final category. With sightings of moa few and far between, this is probably the most important 'type' of cryptozoologist.

The lack of scientific analysis is a weak link in the book, but as I say, this is completely excusable. Because of this, it does not change my opinion that the photograph shows a young red deer. In reality, this is pretty much the only negative point with regards to the three volumes. It did really make me reconsider my initial opinion. However, the data surrounding the picture is unrivalled; the vast majority of it is unavailable on-line as a large quantity of it consists of transcripts of interviews with the witnesses. Sorry internet cryptozoologists... If Spittle published his analysis of the Freaney photograph as a separate, much smaller and more affordable book, along with either a co-author to look scientifically at the photograph, I think he would do the world of cryptozoology a huge favour. These volumes stand alone as a fantastic resource. But, a lot could be done with the data provided. Using Spittle's analysis of the reliability of the reporters, it would be simple to use the remaining information to reject or accept the sightings as being those of a moa, or a deer (or anything else). This book is crying out for extra, more detailed analysis of each sighting.

This book then is exactly how cryptozoology should be done. Has it changed my opinion on the Freaney photograph? No. Has it changed my opinion that there are no large (4ft tall+) species of moa still alive? No, I remain sure that they are extinct. Has it changed my opinion that there are no small species of moa still alive? Sort of I suppose; I think there is a high chance they lived until the 1800's, but I don't think there are any left alive. To finish this review then, I am going to quote Mr. Spittle on why he formed Paua Press Limited in 2007: "Just as a paua [*Haliotis*, a species of New Zealand abalone] appears dull and non-descript on the outside but is of compelling interest when the surface dross is taken away, I am hopeful that the books my press publishes will have, at their centre, something of substance for the reader." He is absolutely correct, this is an incredible book, "*On The Track...*" for moa enthusiasts. For God's sake ask for it as a birthday present... **MB**

Field Guide to Fantastic Creatures
By Giles Sparrow
Hardcover: 144 pages
Publisher: Quercus Publishing Plc
ISBN-10: 184866026X

This is a large format (in fact very large) book dealing with all kinds of monsters, continent by continent. It covers both cryptids like the yeti and giant anaconda, and mythological beasts like the minotaur and medusa. Each creature has a large, colourful computer-art picture, and I have to admit that I am not a big fan of electronically rendered art, preferring proper paintings.

The art varies widely in its execution. Overall, hairy creatures tend to look less convincing when portrayed in this manner, the worst being the yeti, which looks terribly messy and is depicted as being white, a colour that the yeti has never been reported to be. Scaly creatures such as dragons and sea serpents look

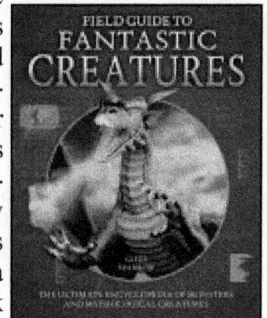

much better. The text is generally quite good, giving information about the creature and maps of its distribution like a real wildlife field guide. Although I don't know him, Giles Sparrow is a member of the CFZ and he was kind enough to mention me in connection with the naga - a giant Indo-Chinese serpent.

This book would make an excellent gift for a younger person and would possibly encourage an interest in the subject for life. **RF**

If Dinosaurs were alive Today
By Dougal Dixon
Hardcover: 96 pages
Publisher: ticktock Media Ltd
ISBN-10: 184696492X

Back in the late 1970s or early '80s (I forget which) Robert Mash wrote a book called *How To Keep Dinosaurs,* written on the basis that were dinosaurs to be alive today some would be kept in zoos and safari parks, others farmed, and others kept as pets. The late Willy Ruston did some cartoons for it.

This book takes the idea somewhat further and shows dinosaurs and other pre-historic animals in the wild and interacting with modern species, which would probably never even have evolved if none of the avian dinosaurs survived. For example, we see *Tyrannosaurus rex,* hunting herds of steers on the plains of Texas, and a Styracosaurus facing off against a white rhino in South Africa. Computer rendered creatures are slotted quite well into photographs. We do see the oft repeated mistake originating in the series *Walking With Dinosaurs* that the pliosaur Liopleurodon was 24 metres long and weighed 90 tons. As far as we know, it grew to around 9 metres long, though other species were larger.

Once again a nice book for younger readers. **RF**

The Butterfly Isles: A Summer In Search Of Our Emperors And Admirals by Patrick Barkham
Paperback: 304 pages
Publisher: Granta Books (5 May 2011)
Language English
ISBN-10: 1847083153

As readers of my inky fingered scribblings both here and elsewhere will know, I have an immense fondess for chatty lepidoptera memoirs of the type popularised by Philip Allan and L Hugh Newman. It is a genre which doesn't really exist any more, so I was overjoyed to find Barkham's book. It reads like a naturalist's memoir from the early part of the 20th Century but it includes such contemporary subjects as text messages, traffic jams and unsatisfactory girlfriends. It includes such wonderful snippets of information as the best way to attract purple emperors (they have an inordinate fondness for a disgusting fish paste from Ghana with the eminently suitable name of Shito) a swarm of heath fritilaries in Kent, which is probably unprecedented in the last century, and even a John Belushi joke.

This is the nicest book that I have read in years, and even has some uses for the cryptozoologists such as me who find the advent of such incredibaly rare migrants such as the Queen of Spain fritilary to be something of quaisi-cryptozoological import. You must buy this book.. I insist. **JD**

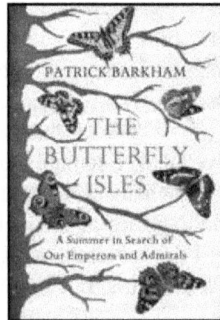

Killers on the Moor
by Mike Freebury
Hardcover: 400 pages
Publisher: Book Guild Ltd
Language English
ISBN-10: 1846245842

This book is very well produced, and Freebury is an engaging, likeable and witty writer, and his tale of investigating animal mutilations at home and abroad is well written and enjoyable.

However, and this is such a big however that it needs to be printed: HOWEVER

There are some unforgiveable mistakes. Alarm bells began ringing in my mind as soon as I started reading his account of the Newquay Zoo animal mutilations of 1977, and more specifically the involvement of the late Mrs Joan Amos in investigating them.

I knew Joan quite well, and I am certain that she told me that she had not investigated the matter for herself, but had only received the documents and information third hand via a contact in the Plymouth UFO Group. I published a lengthy account of the case in *The Owlman and Others* (first published 1997), in which I put at least one of the outstanding queries to bed.

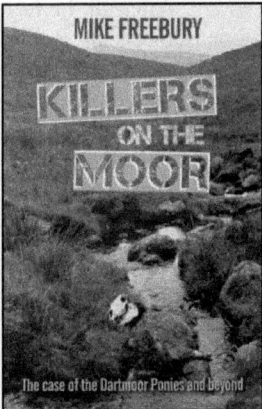

The lumps on the jaws of the wallabies were due to an infection by *Fusobacterium necrophorous*, a condition known colloquially as 'lumpy jaw.' It is a well-known condition in captive marsupials and not at all mysterious or sinister.

I cannot prove that Freebury is wrong about Joan's involvement in the case because she has been dead for years and it is my memory versus his word! However, when he comes on to the next case, the Loftus wallaby slashing, I can and will provide a list of refutations:

1. The case did not take place in 2003. It was August 2002; the 21-23rd to be exact (my 43rd birthday).
2. I have never been in charge of an organisation called the Crypto Zoologist Foundation. Presumably he means the Centre for Fortean Zoology (CFZ)
3. I did not send Richard to Loftus. The two of us travelled up there together.
4. The team from *Scream Team* were not in the area by chance. I arranged to liaise with them, and they paid Richard's and my expenses.

...and so on.

I would like to think that these mistakes are just one-off aberrations.

The book is otherwise very interesting and makes some good points. However, in a later chapter the descriptions of the events surrounding Joan's time in hospital during 1978 are at variance with my memory of what she told me, and there are large chunks of her research into animal mutilations and missing domestic cats that have been ignored, so I am inclined to think that other parts of the book may be equally as flawed.

So this is strange. I like the book, and I agree with many of his conclusions, but what the bloody hell was he thinking when he was researching it? **JD**

Deep in a cave beneath Loch Ness lives a strange figure who steals ideas from other magazines and then somehow makes them his own.

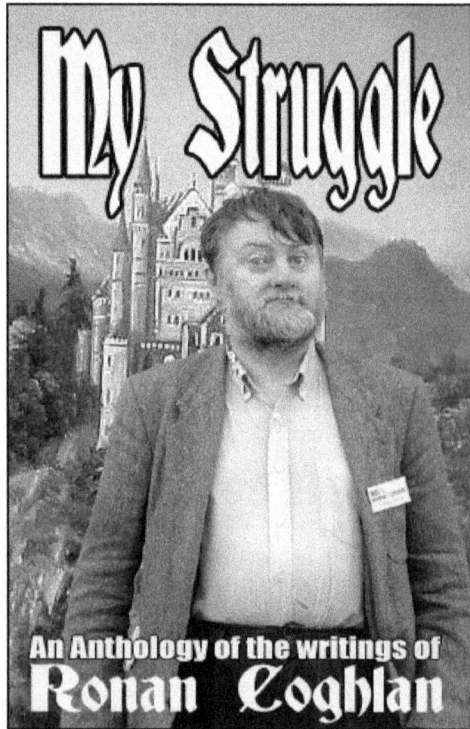

My Struggle

An Anthology of the writings of
Ronan Coghlan

RONAN - HIS LIFE AND LOVES

What can one say about Ronan Coghlan? He has been a fixture at the Weird Weekend for the last seven years (for each of which Richard Freeman has introduced him as a 'feckless rogue'.) He really is a most singular fellow. This year's extravaganza, during which he demonstrated the Twelve Labours of Hercules with the aid of two small children and a fearsome-looking machine gun that squirted water was unprecedented, but his contribution to the CFZ Anthology series was simply jaw-dropping.

As regular readers will know, CFZ Press has been publishing anthologies of the work of various Fortean writers (Andy Roberts, Nick Redfern and Paul Screeton to date) and it seemed like a jolly good idea to invite the 'feckless rogue' himself to contribute a volume to this august series. After all, he has written on such esoteric subjects as Robin Hood, the wee folk, cryptozooology and Sherlock Holmes (a peculiar novel in which the giant rat of Sumatra takes human shape, and various CFZ luminaries including our beloved Herr Obergrupenfuhrer are very vaguely lampooned). They were expecting a collection of such scholarly vignettes, but

THE SYCOPHANT

what they actually got was a 15,000 word autobiography which featured such astounding events from the Coghlan life and loves as being kidnapped by Chinese pirates.

The jury is still out regarding what the hell will be done with this remarkable document, although a cover has been prepared by the everdiligent CFZ art department (Jon).

FAMOUS BLUE RAINCOAT

People across the interwebs seem to delight in jumping upon various CFZ memories as soon as they can get away with accusing them of any one of a string of crimes against morality, the state, or nature (delete where applicable). We were surprised, therefore, when - early this year - Richard Freeman and Adam Davies made a public appeal for long raincoats to be taken to Sumatra as gifts for the guides that this did not provoke a tirade of flasher jokes. After all, as Rod Steward said, an old raincoat will never let you down.

But they singularly failed to appear. Perhaps The Sycophant has a dirtier mind, and a more prurient soul than other CFZ-watchers out there in cyberspace.

IT MAKES ME FEEL GREATER

Each year the Weird Weekend think-tank is faced with a seemingly insurmountable problem, namely how do they surpass the previous year's bout of silliness. They have various criteria which they have to fill—it must feature schoolkids (or preferably even younger) sing-

ing a song by *Hawkwind* or some other scion of the British underground music scene of the early 1970s, and it has to end up with Melinda Harding (a young lady of the village) being chased around by a dragon, an alien, or (as was the case in 2010) some frog goblins.

As can be seen by this remarkable picture Bo (aged six) on vocals ably assisted by Prudence, her sister Ruby, another young lady called Kelsey, and Graham on synthesiser, provided a more than adequate rendition of *Orgone Accumulator* which ended up with Richard and Melinda being chased around the Community Centre by a dragon.

Rumours that next year the tiny tots will be singing *Let's loot the supermarket (like we did last summer)* by *Mick Farren and The Deviants* cannot be confirmed at the time of going to press.

STILL ON THE TRACK OF UNKNOWN ANIMALS

The Centre for Fortean Zoology, or CFZ, is a non profit-making organisation founded in 1992 with the aim of being a clearing house for information, and coordinating research into mystery animals around the world.

We also study out of place animals, rare and aberrant animal behaviour, and Zooform Phenomena; little-understood "things" that appear to be animals, but which are in fact nothing of the sort, and not even alive (at least in the way we understand the term).

Not only are we the biggest organisation of our type in the world, but - or so we like to think - we are the best. We are certainly the only truly global cryptozoological research organisation, and we carry out our investigations using a strictly scientific set of guidelines. We are expanding all the time and looking to recruit new members to help us in our research into mysterious animals and strange creatures across the globe.

Why should you join us? Because, if you are genuinely interested in trying to solve the last great mysteries of Mother Nature, there is nobody better than us with whom to do it.

Members get a four-issue subscription to our journal *Animals & Men*. Each issue contains nearly 100 pages packed with news, articles, letters, research papers, field reports, and even a gossip column! The magazine is Royal Octavo in format with a full colour cover. You also have access to one of the world's largest collections of resource material dealing with cryptozoology and allied disciplines, and people from the CFZ membership regularly take part in fieldwork and expeditions around the world.

The CFZ is managed by a three-man board of trustees, with a non-profit making trust registered with HM Government Stamp Office. The board of trustees is supported by a Permanent Directorate of full and part-time staff, and advised by a Consultancy Board of specialists - many of whom are world-renowned experts in their particular field. We have regional representatives across the UK, the USA, and many other parts of the world, and are affiliated with other organisations whose aims and protocols mirror our own.

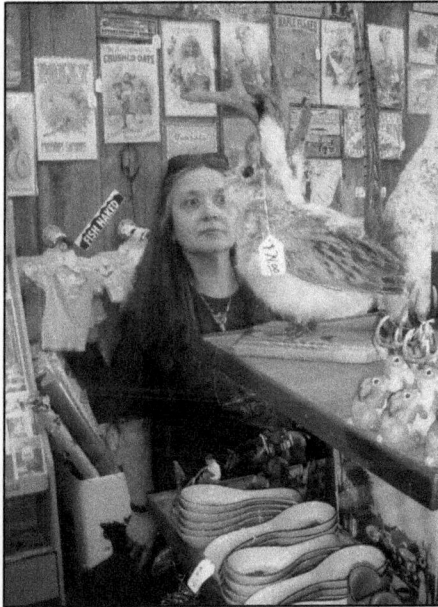

You'll find that the people at the CFZ are friendly and approachable. We have a thriving forum on the website which is the hub of an ever-growing electronic community. You will soon find your feet. Many members of the CFZ Permanent Directorate started off as ordinary members, and now work full-time chasing monsters around the world.

Write to us, e-mail us, or telephone us. The list of future projects on the website is not exhaustive. If you have a good idea for an investigation, please tell us. We may well be able to help.

We are always looking for volunteers to join us. If you see a project that interests you, do not hesitate to get in touch with us. Under certain circumstances we can help provide funding for your trip. If you look on the future projects section of the website, you can see some of the projects that we have pencilled in for the next few years.

In 2003 and 2004 we sent three-man expeditions to Sumatra looking for Orang-Pendek - a semi-legendary bipedal ape. The same three went to Mongolia in 2005. All three members started off merely subscribers to the CFZ magazine. Next time it could be you!

We have no magic sources of income. All our funds come from donations, membership fees, and sales of our publications and merchandise. We are always looking for corporate sponsorship, and other sources of revenue. If you have any ideas for fund-raising please let us know.

However, unlike other cryptozoological organisations in the past, we do not live in an intellectual ivory tower. We are not afraid to get our hands dirty, and furthermore we are not one of those organisations where the membership have to raise money so that a privileged few can go on expensive foreign trips. Our research teams, both in the UK and abroad, consist of a mixture of experienced and inexperienced personnel. We are truly a community, and work on the premise that the benefits of CFZ membership are open to all.

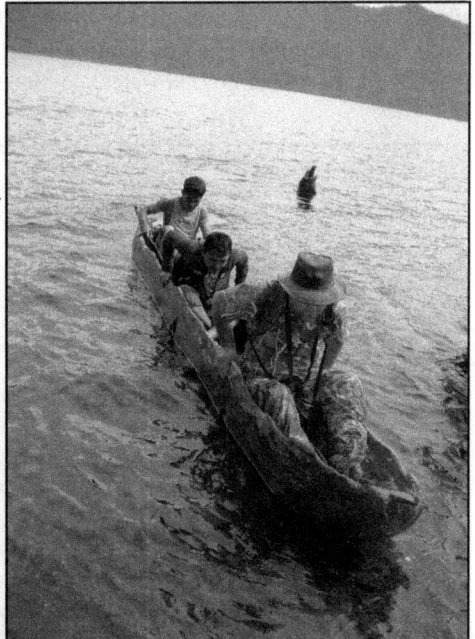

Reports of our investigations are published on our website as soon as they are available. Preliminary reports are posted within days of the project finishing.

Each year we publish a 200 page yearbook containing research papers and expedition reports too long to be printed in the journal. We freely circulate our information to anybody who asks for it.

We have a thriving YouTube channel, CFZtv, which has well over two hundred self-made documentaries, lecture appearances, and episodes of our monthly webTV show. We have a daily online magazine, which has over a million hits each year.

Each year since 2000 we have held our annual convention - the Weird Weekend. It is three days of lectures, workshops, and excursions. But most importantly it is a chance for members of the CFZ to meet each other, and to talk with the members of the permanent directorate in a relaxed and informal setting and preferably with a pint of beer in one hand. Since 2006 - the Weird Weekend has been bigger and better and held on the third weekend in August in the idyllic rural location of Woolsery in North Devon.

Since relocating to North Devon in 2005 we have become ever more closely involved with other community organisations, and we hope that this trend will continue. We have also worked closely with Police Forces across the UK as consultants for animal mutilation cases, and we intend to forge closer links with the coastguard and other community services. We want to work closely with those who regularly travel into the Bristol Channel, so that if the recent trend of exotic animal visitors to our coastal waters continues, we can be out there as soon as possible.

Apart from having been the only Fortean Zoological organisation in the world to have consistently published material on all aspects of the subject for over a decade, we have achieved the following concrete results:

• Disproved the myth relating to the headless so-called sea-serpent carcass of Durgan beach in Cornwall 1975
• Disproved the story of the 1988 puma skull of

Lustleigh Cleave

- Carried out the only in-depth research ever into the mythos of the Cornish Owlman.
- Made the first records of a tropical species of lamprey
- Made the first records of a luminous cave gnat larva in Thailand
- Discovered a possible new species of British mammal - the beech marten
- In 1994-6 carried out the first archival fortean zoological survey of Hong Kong
- In the year 2000, CFZ theories were confirmed when a new species of lizard was added to the British List
- Identified the monster of Martin Mere in Lancashire as a giant wels catfish
- Expanded the known range of Armitage's skink in the Gambia by 80%
- Obtained photographic evidence of the remains of Europe's largest known pike
- Carried out the first ever in-depth study of the ninki-nanka
- Carried out the first attempt to breed Puerto Rican cave snails in captivity
- Were the first European explorers to visit the `lost valley` in Sumatra
- Published the first ever evidence for a new tribe of pygmies in Guyana
- Published the first evidence for a new species of caiman in Guyana
- Filmed unknown creatures

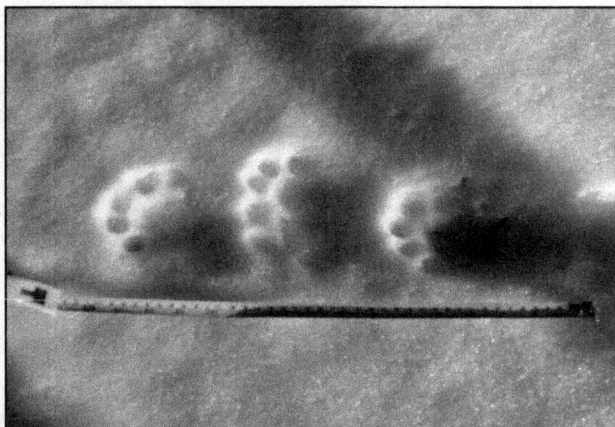

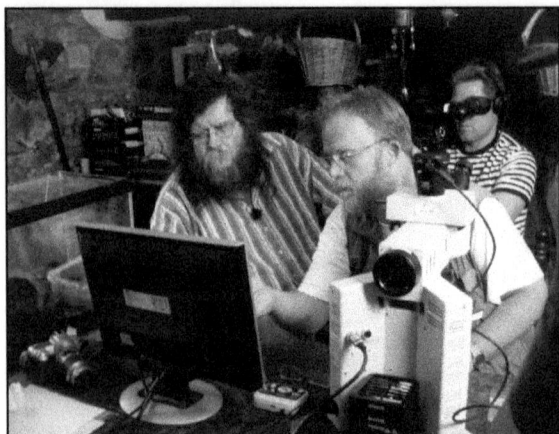

on a monster-haunted lake in Ireland for the first time

• Had a sighting of orang pendek in Sumatra in 2009

• Found leopard hair, subsequently identified by DNA analysis, from rural North Devon in 2010

• Brought back hairs which appear to be from an unknown primate in Sumatra

• Published some of the best evidence ever for the almasty in southern Russia

CFZ Expeditions and Investigations include:

• 1998 Puerto Rico, Florida, Mexico (Chupacabras)
• 1999 Nevada (Bigfoot)
• 2000 Thailand (Naga)
• 2002 Martin Mere (Giant catfish)
• 2002 Cleveland (Wallaby mutilation)

- 2003 Bolam Lake (BHM Reports)
- 2003 Sumatra (Orang Pendek)
- 2003 Texas (Bigfoot; giant snapping turtles)
- 2004 Sumatra (Orang Pendek; cigau, a sabre-toothed cat)
- 2004 Illinois (Black panthers; cicada swarm)
- 2004 Texas (Mystery blue dog)
- Loch Morar (Monster)
- 2004 Puerto Rico (Chupacabras; carnivorous cave snails)
- 2005 Belize (Affiliate expedition for hairy dwarfs)
- 2005 Loch Ness (Monster)
- 2005 Mongolia (Allghoi Khorkhoi aka Mongolian death worm)

- 2006 Gambia (Gambo - Gambian sea monster , Ninki Nanka and Armitage's skink
- 2006 Llangorse Lake (Giant pike, giant eels)
- 2006 Windermere (Giant eels)
- 2007 Coniston Water (Giant eels)
- 2007 Guyana (Giant anaconda, didi, water tiger)
- 2008 Russia (Almasty)
- 2009 Sumatra (Orang pendek)
- 2009 Republic of Ireland (Lake Monster)
- 2010 Texas (Blue Dogs)
- 2010 India (Mande Burung)
- 2011 Sumatra (Orang pendek)

For details of current membership fees, current expeditions and investigations, and voluntary posts within the CFZ that need your help, please do not hesitate to contact us.

The Centre for Fortean Zoology,
Myrtle Cottage,
Woolfardisworthy,
Bideford, North Devon
EX39 5QR

Telephone 01237 431413
Fax+44 (0)7006-074-925
eMail info@cfz.org.uk

Websites:

www.cfz.org.uk
www.weirdweekend.org

HOW TO START A PUBLISHING EMPIRE

Unlike most mainstream publishers, we have a non-commercial remit, and our mission statement claims that "we publish books because they deserve to be published, not because we think that we can make money out of them". Our motto is the Latin Tag *Pro bona causa facimus* (we do it for good reason), a slogan taken from a children's book *The Case of the Silver Egg* by the late Desmond Skirrow.

WIKIPEDIA: "The first book published was in 1988. *Take this Brother may it Serve you Well* was a guide to Beatles bootlegs by Jonathan Downes. It sold quite well, but was hampered by very poor production values, being photocopied, and held together by a plastic clip binder. In 1988 A5 clip binders were hard to get hold of, so the publishers took A4 binders and cut them in half with a hacksaw. It now reaches surprisingly high prices second hand.

The production quality improved slightly over the years, and after 1999 all the books produced were ringbound with laminated colour covers. In 2004, however, they signed an agreement with Lightning Source, and all books are now produced perfect bound, with full colour covers."

Until 2010 all our books, the majority of which are/were on the subject of mystery animals and allied disciplines, were published by `CFZ Press`, the publishing arm of the Centre for Fortean Zoology (CFZ), and we urged our readers and followers to draw a discreet veil over the books that we published that were completely off topic to the CFZ.

However, in 2010 we decided that enough was enough and launched a second imprint, `Fortean Words` which aims to cover a wide range of non animal-related esoteric subjects. Other imprints will be launched as and when we feel like it, however the basic ethos of the company remains the same: Our job is to publish books and magazines that we feel are worth publishing, whether or not they are going to sell. Money is, after all - as my dear old Mama once told me - a rather vulgar subject, and she would be rolling in her grave if she thought that her eldest son was somehow in `trade`.

Luckily, so far our tastes have turned out not to be that rarified after all, and we have sold far more books than anyone ever thought that we would, so there is a moral in there somewhere…

Jon Downes,
Woolsery, North Devon
July 2010

CFZ PRESS

Other Books in Print

Weird Waters – The Mystery Animals of Scandinavia: Lake and Sea Monsters by Lars Thomas
The Inhumanoids by Barton Nunnelly
Monstrum! A Wizard's Tale by Tony "Doc" Shiels
CFZ Yearbook 2011 edited by Jonathan Downes
Karl Shuker's Alien Zoo by Shuker, Dr Karl P.N
Tetrapod Zoology Book One by Naish, Dr Darren
The Mystery Animals of Ireland by Gary Cunningham and Ronan Coghlan
Monsters of Texas by Gerhard, Ken
The Great Yokai Encyclopaedia by Freeman, Richard
NEW HORIZONS: Animals & Men *issues 16-20 Collected Editions Vol. 4*
by Downes, Jonathan
A Daintree Diary -
Tales from Travels to the Daintree Rainforest in tropical north Queensland, Australia
by Portman, Carl
Strangely Strange but Oddly Normal by Roberts, Andy
Centre for Fortean Zoology Yearbook 2010 by Downes, Jonathan
Predator Deathmatch by Molloy, Nick
Star Steeds and other Dreams by Shuker, Karl
CHINA: A Yellow Peril? by Muirhead, Richard
Mystery Animals of the British Isles: The Western Isles by Vaudrey, Glen
Giant Snakes - Unravelling the coils of mystery by Newton, Michael
Mystery Animals of the British Isles: Kent by Arnold, Neil
Centre for Fortean Zoology Yearbook 2009 by Downes, Jonathan
CFZ EXPEDITION REPORT: Russia 2008 by Richard Freeman *et al*, Shuker, Karl (fwd)
Dinosaurs and other Prehistoric Animals on Stamps - A Worldwide catalogue
by Shuker, Karl P. N
Dr Shuker's Casebook by Shuker, Karl P.N
The Island of Paradise - chupacabra UFO crash retrievals,
and accelerated evolution on the island of Puerto Rico by Downes, Jonathan
The Mystery Animals of the British Isles: Northumberland and Tyneside by Hallowell, Michael J
Centre for Fortean Zoology Yearbook 1997 by Downes, Jonathan (Ed)
Centre for Fortean Zoology Yearbook 2002 by Downes, Jonathan (Ed)
Centre for Fortean Zoology Yearbook 2000/1 by Downes, Jonathan (Ed)

Centre for Fortean Zoology Yearbook 1998 by Downes, Jonathan (Ed)
Centre for Fortean Zoology Yearbook 2003 by Downes, Jonathan (Ed)
In the wake of Bernard Heuvelmans by Woodley, Michael A
CFZ EXPEDITION REPORT: Guyana 2007 by Richard Freeman *et al*, Shuker, Karl (fwd)
Centre for Fortean Zoology Yearbook 1999 by Downes, Jonathan (Ed)
Big Cats in Britain Yearbook 2008 by Fraser, Mark (Ed)
Centre for Fortean Zoology Yearbook 1996 by Downes, Jonathan (Ed)
THE CALL OF THE WILD - Animals & Men issues 11-15
Collected Editions Vol. 3 by Downes, Jonathan (ed)
Ethna's Journal by Downes, C N
Centre for Fortean Zoology Yearbook 2008 by Downes, J (Ed)
DARK DORSET -Calendar Custome by Newland, Robert J
Extraordinary Animals Revisited by Shuker, Karl
MAN-MONKEY - In Search of the British Bigfoot by Redfern, Nick
Dark Dorset Tales of Mystery, Wonder and Terror by Newland, Robert J and Mark North
Big Cats Loose in Britain by Matthews, Marcus
MONSTER! - The A-Z of Zooform Phenomena by Arnold, Neil
The Centre for Fortean Zoology 2004 Yearbook by Downes, Jonathan (Ed)
The Centre for Fortean Zoology 2007 Yearbook by Downes, Jonathan (Ed)
CAT FLAPS! Northern Mystery Cats by Roberts, Andy
Big Cats in Britain Yearbook 2007 by Fraser, Mark (Ed)
BIG BIRD! - Modern sightings of Flying Monsters by Gerhard, Ken
THE NUMBER OF THE BEAST - Animals & Men issues 6-10
Collected Editions Vol. 1 by Downes, Jonathan (Ed)
IN THE BEGINNING - Animals & Men *issues 1-5 Collected Editions Vol. 1* by Downes, Jonathan
STRENGTH THROUGH KOI - They saved Hitler's Koi and other stories by Downes, Jonathan
The Smaller Mystery Carnivores of the Westcountry by Downes, Jonathan
CFZ EXPEDITION REPORT: Gambia 2006 by Richard Freeman *et al*, Shuker, Karl (fwd)
The Owlman and Others by Jonathan Downes
The Blackdown Mystery by Downes, Jonathan
Big Cats in Britain Yearbook 2006 by Fraser, Mark (Ed)
Fragrant Harbours - Distant Rivers by Downes, John T
Only Fools and Goatsuckers by Downes, Jonathan
Monster of the Mere by Jonathan Downes
Dragons:More than a Myth by Freeman, Richard Alan
Granfer's Bible Stories by Downes, John Tweddell
Monster Hunter by Downes, Jonathan

Fortean Words

The Centre for Fortean Zoology has for several years led the field in Fortean publishing. CFZ Press is the only publishing company specialising in books on monsters and mystery animals. CFZ Press has published more books on this subject than any other company in history and has attracted such well known authors as Andy Roberts, Nick Redfern, Michael Newton, Dr Karl Shuker, Neil Arnold, Dr Darren Naish, Jon Downes, Ken Gerhard and Richard Freeman.

Now CFZ Press are launching a new imprint. Fortean Words is a new line of books dealing with Fortean subjects other than cryptozoology, which is - after all - the subject the CFZ are best known for. Fortean Words is being launched with a spectacular multi-volume series called *Haunted Skies* which covers British UFO sightings between 1940 and 2010. Former policeman John Hanson and his long-suffering partner Dawn Holloway have compiled a peerless library of sighting reports, many that have not been made public before.

Other books include a look at the Berwyn Mountains UFO case by renowned Fortean Andy Roberts and a series of forthcoming books by transatlantic researcher Nick Redfern. CFZ Press are dedicated to maintaining the fine quality of their works with Fortean Words. New authors tackling new subjects will always be encouraged, and we hope that our books will continue to be as ground-breaking and popular as ever.

Haunted Skies Volume One 1940-1959 by John Hanson and Dawn Holloway
Haunted Skies Volume Two 1960-1965 by John Hanson and Dawn Holloway
Space Girl Dead on Spaghetti Junction - an anthology by Nick Redfern
I Fort the Lore - an anthology by Paul Screeton
UFO Down - the Berwyn Mountains UFO Crash by Andy Roberts